REMOTE SENSING
OF SNOW AND ICE

REMOTE SENSING OF SNOW AND ICE

W. Gareth Rees

Cambridge University
England

CRC Press
Taylor & Francis Group
Boca Raton London New York

CRC Press is an imprint of the
Taylor & Francis Group, an **informa** business

CRC Press
Taylor & Francis Group
6000 Broken Sound Parkway NW, Suite 300
Boca Raton, FL 33487-2742

First issued in paperback 2019

© 2006 by Taylor & Francis Group, LLC
CRC Press is an imprint of Taylor & Francis Group, an Informa business

No claim to original U.S. Government works

ISBN-13: 978-0-415-29831-5 (hbk)
ISBN-13: 978-0-367-39230-7 (pbk)

Library of Congress Card Number 2005048616

Library of Congress Cataloging-in-Publication Data

Rees, Gareth, 1959-
    Remote sensing of snow and ice / by W. Gareth Rees.
        p. cm.
    Includes bibliographical references and index.
    ISBN 0-415-29831-8 (alk. paper)
    1. Snow--Remote sensing. 2. Ice--Remote sensing. I. Title.

GB2601.72.R42R44 2005
551.57'84'0285--dc22   2005048616

**Visit the Taylor & Francis Web site at
http://www.taylorandfrancis.com**

**and the CRC Press Web site at
http://www.crcpress.com**

Dedication

———————

To the memory of my father, 1933–2003

Preface

Although our planet is characterized by water, the extent to which water in its frozen state plays a role in the functioning of the Earth is perhaps under-appreciated. Snow and ice cover about a sixth of the Earth's surface, as snow lying on the ground, as glaciers and larger masses of terrestrial ice, including the huge ice sheets of Antarctica and Greenland, as sea ice, river and lake ice, and as icebergs. Together with the subsurface ice of permafrost, these forms of ice constitute the Earth's cryosphere. The cryosphere plays a significant function in controlling the global climate system, and has an obvious role as an indicator of change in the system, but aspects of it are important at all spatial scales, to a wide range of topics that includes ecology, hydrology, economics, transport, and recreation.

Most of the Earth's snow and ice is located in remote places where access is difficult, climate is extreme, and the polar night complicates the task of collecting information. For these reasons, and because of the enormous areas represented by snow and ice, the observational techniques of remote sensing from the air and especially from satellites are particularly useful as a means of gathering data about the cryosphere. These techniques offer the possibility of collecting large amounts of information in a short time. Rates of viewing the Earth's surface can be as high as $100,000 \, km^2$ per second for some spaceborne systems, easily capable of giving a synoptic view of large parts of the Earth's surface, although the most useful systems have superficial rates typically a factor of 10 or 100 less than this. Some remote sensing systems can view the Earth's surface through cloud cover; some are unaffected by the absence of daylight and so are able to function during the polar night. Virtually all remote sensing systems now produce digital data, which are suitable for storage and analysis by computer. These virtues were recognized early by researchers concerned with snow and ice. Since the emergence of spaceborne technologies for monitoring the cryosphere (around 1970), new methods have been developed, the number of sensors that are used simultaneously has increased, and progressively longer runs of self-consistent data have become available.

The aim of this book is to provide a reasonably comprehensive introduction to the remote sensing of the Earth's cryosphere. It should provide enough background information to explain why cryospheric observations are important and why remote sensing observations are desirable or even essential, and an outline of remote sensing techniques and methods in general, but the principal areas that it covers are (1) a comprehensive survey of the essential physical properties of ice and snow, and (2) an up-to-date survey of the existing remote sensing techniques and those being developed. Specifically, the book discusses snow cover, "land ice" (glaciers, ice sheets, and ice shelves), sea, river, and lake ice, and icebergs. I have envisaged the book as a successor to *Remote*

Sensing of Ice and Snow (Hall and Martinec, Chapman & Hall, 1985), and the title is deliberately similar. I have been recommending Hall and Martinec's book to my students for many years, but there have been major advances in techniques of remote sensing and data analysis since it was published, and much more has been learnt about the cryosphere. This book tries to reflect these advances. Of course, much has been published in this area since 1985, but it has been largely confined to research journals (notably, though far from exclusively, the issues of *Annals of Glaciology* presenting the proceedings of the International Glaciological Society's occasional international symposia on Remote Sensing in Glaciology) and edited volumes treating some particular aspect of the field. Neither of these approaches provides the uniformity of treatment, or the possibility of identifying unifying themes and ideas, that is possible in a single volume. Ultimately, my motivation for wanting to write this book is my frustration that nothing like it really exists at present.

It is not obvious how to construct a book with this title. Should the author assume that readers are fully familiar with the cryosphere but know nothing about the techniques of remote sensing and image analysis, or conversely, that they are remote sensing specialists who know nothing about the cryosphere? Should the book be a compendium of recent research results? None of these approaches seems ideal. Rather, the book assumes the reader to have a limited knowledge of both the cryosphere, or at least some aspects of it, and of remote sensing methods in general, and this assumption has informed its structure and scope. The first chapter presents a brief review of the nature and significance of the cryosphere, while Chapter 2 performs a similar function for the principal techniques of remote sensing, including surveys of the main spaceborne instruments. Collection of data about a region of the Earth's surface is of limited use without subsequent analysis, thus Chapter 3 presents an overview of methods of digital image analysis. The emphasis in this chapter is on methods that are currently used or being developed for specific cryospheric applications, and a number of these methods are amplified in the later parts of the book, where they are described in context. From this point, the book returns to the subject of snow and ice. Chapter 4 is a reasonably technical review of the physical properties of snow and ice as they relate to remote sensing observations. It is thus concerned with the principles by which useful geophysical information becomes encoded into the electromagnetic radiation that is detected by the remote sensing process, and follows fairly directly from Chapter 1. The next five chapters discuss in detail the applications of remote sensing methods to the study of different cryospheric components: snow (Chapter 5), sea ice (Chapter 6), freshwater ice (Chapter 7), glaciers (Chapter 8), and icebergs (Chapter 9). Permafrost is not treated in this part of the book, for reasons that are discussed in Chapter 1. The final chapter is a summary of this section of the book, discussing what remote sensing has revealed about the cryosphere, where the main technical problems exist, and the scope for addressing these problems. In terms of balance, about a quarter of the book is devoted to general aspects of the cryosphere, a quarter to introducing remote sensing and image processing methods, and the remaining half to detailed discussion of the application of remote sensing methods to particular components of the cryosphere.

The book has been written for research workers in the environmental sciences generally and in the cryosphere in particular, though it should also be accessible to students at the master's degree and senior undergraduate level. Although it does not set out to present a compendium of recent research results, the book is extensively referenced with about 650 citations to the recent scientific literature, and it is hoped that, although this list cannot claim to be comprehensive (and I apologise in advance for the inevitable omission of important references), it will allow the reader who wishes to do so to become familiar with the latest research in a particular field. The treatment throughout the book is moderately technical, although Chapter 4, which discusses the physical properties of ice and snow and how they relate to remote sensing observations, is necessarily a little more mathematical in places. It is not necessary to absorb all of this mathematical detail to understand the rest of the book.

It is my intention that this book should function in part as a work of reference. A good index is highly desirable in such a work since it facilitates cross-referencing. The index to this book has been constructed with this principle in mind, and it includes the names of satellites and their sensors for easy location of examples of applications making use of a particular remote sensing instrument. It also includes geographical names, so that examples from a particular glacier, such as Midre Lovénbreen on Svalbard, can be found. These examples are also indexed under "Svalbard."

W. G. Rees
Cambridge

Acknowledgments

As always, my first debt is to my wife, Christine, for her encouragement and support in many ways. I also thank Dr. Neil Arnold of the Scott Polar Research Institute, University of Cambridge, for research collaboration and many discussions, in Svalbard and in Cambridge. The Taylor & Francis Group has been patient and helpful.

The following people and organizations have given permission for the use of data and illustrations, which is gratefully acknowledged: American Geophysical Union (Figures 6.8, 6.11, 6.12, 6.16, 9.6); the Arctic Institute of North America (Figures 7.3 to 7.5); Elsevier Ltd (Figure 4.14); Cold Regions Research and Engineering Laboratory (Figure 7.8); the European Association of Remote Sensing Laboratories (Figure 4.3); the European Space Agency (Figures 2.23, 2.24, 5.3, 6.6, 8.5); Richard Flanders, Toolik Field Station, Alaska (Figure 1.20); Geological Survey of Canada (Figure 1.10); Global Land Ice Measurements from Space Canadian Regional Center, University of Alberta (Figure 1.8); Faye Hicks, Department of Civil and Environmental Engineering, University of Alberta (Figure 1.19); J. Dana Hrubes, South Pole Station, Antarctica (Figure 1.17); IEEE (Figure 8.11); International Glaciological Society (Figures 5.6, 6.7, 6.14, 8.2, 8.6 to 8.10, 8.12 to 8.14); International Society of Offshore and Polar Engineering (Figure 9.4); John Lin, University of Cambridge (Figure 2.24); Nansen Environmental and Remote Sensing Center, Bergen, Norway (Figure 6.6); NASA Earth Observatory (Figure 1.7); NASA Visible Earth (Figure 5.1); National Snow and Ice Data Center, University of Colorado, Boulder (Figures 1.2, 1.14, 1.15, 5.5); NERC Satellite Receiving Station, University of Dundee (Figure 2.7); NPO Planeta, Moscow (Figure 6.6); Polar Record (Figures 7.6, 7.7); Rutherford Appleton Laboratory (Figure 2.9); Taylor & Francis (Figures 6.1, 6.13, 6.15, 6.18, 9.3, 9.5); Tromsø Satellite Station, Norway (Figure 6.6); University of Leicester Earth Observation Science Group (Figure 2.9); U.S. Coast Guard International Ice Patrol (Figure 1.13); Richard Waller, Keele University (Figure 1.6).

Author Biography

Dr. Rees read Natural Sciences at the University of Cambridge, and completed his PhD in radio astronomy at the Cavendish Laboratory. Since 1985 he has been head of the Remote Sensing Group at the Scott Polar Research Institute, University of Cambridge.

His current research interests include remote sensing of glaciers, snow cover and high-latitude vegetation. He has conducted frequent fieldwork visits to the Arctic, and is the author of 5 books and over 80 scientific papers.

Table of Contents

1 The Cryosphere

1.1 INTRODUCTION

The Earth's snowy and icy regions are collectively known as the *cryosphere*, a word derived from the Greek *krios* meaning "cold." The constituents of the cryosphere are thus the snow cover, sea ice, freshwater ice (frozen lakes and rivers), the large ice masses on land (ice sheets, glaciers, and related phenomena such as ice shelves and icebergs), and permafrost. Temperatures on the Earth's surface generally decrease with increasing distance from the equator, so the cryosphere is largely a high-latitude phenomenon (see Color Figure 1.1 following page 108).

The presence of ice and snow on the Earth is significant over a wide range of spatial and temporal scales. Locally and regionally, the cryosphere interacts with the human and natural environments in both positive and negative ways. On the global scale, the cryosphere represents an important part of the Earth's climate system. Ice and snow are generally highly reflective of incident (shortwave) solar radiation, so they provide a feedback mechanism to the system (crudely, more cryosphere means that the Earth retains less of the incident solar radiation). Their presence, and annual and long-term variations, modify the distribution and flow of water. Models of global climate change predict that the largest changes, generally speaking, will occur at high latitudes (the so-called *polar amplification*), so it is important to monitor the polar regions generally, and the cryosphere in particular, to look for manifestations of global climate change. The remoteness of much of the cryosphere from centers of population, and its hostile environment, means that investigation by means of remote sensing methods, particularly based on the use of satellite data, is desirable (Derksen et al. 2002; Massom 1991), and indeed the cryosphere has been monitored from space since the mid-1960s (Foster and Chang 1993).

This chapter describes the main components of the cryosphere, their spatial distribution, importance, and the need and scope for monitoring them.

1.2 SNOW COVER

Snow is defined as falling or deposited ice particles formed mainly by sublimation (Unesco/IAHS/WMO 1970). In this book we are concerned with *snow cover*, rather than falling snow (Figure 1.1). Roughly 5% of the global precipitation that reaches the Earth's surface does so in the form of snow

1

FIGURE 1.1 Snow cover in the Scottish highlands (author's photograph).

(Hoinkes 1967), although in the Arctic regions this proportion reaches 50 to 90% (Winther and Hall 1999). It is convenient to distinguish between *permanent*, *seasonal*, and *temporary* snow cover. Temporary and seasonal snow covers do not survive the summer, while permanent snow cover is retained for many years. The latter occurs principally in Antarctica and Greenland, so that the great majority of permanent snow cover is a southern hemisphere phenomenon, while temporary and seasonal snow cover* are predominantly northern hemisphere phenomena. Roughly speaking, the distribution of temporary and seasonal snow can be characterized as follows. In North America, temporary snow occurs between latitudes 30 and 40°, while seasonal snow occurs north of 40°. In western Europe, seasonal snow occurs north of latitude 60° and in mountainous areas, while temporary snow can occur almost anywhere except in the south-west of the Iberian Peninsula. In eastern Europe, seasonal snow extends northward from about 50° latitude and temporary snow reaches as far south as latitude 35° in the Middle East. Seasonal snow in Asia occurs as far south as 30° latitude. In the southern hemisphere, most snow cover is confined to mountainous areas (the Andes, Drakensberg, Snowy Mountains, and the Southern Alps of New Zealand). The global distribution of snow cover is summarized in Figure 1.2.

The spatial extent of the northern hemisphere snow cover, excluding Greenland, varies between about 4 million square kilometers in August and 46 million square kilometers (roughly 40% of the land area) (Frei and Robinson 1999) in January (Figure 1.3). The maximum mass of water represented by this

---

*Temporary and seasonal snow are differentiated according to the length of time for which they persist. Typically, a seasonal snow cover will survive for several months, usually being replenished throughout the winter, while a temporary snow cover survives for a matter of days.

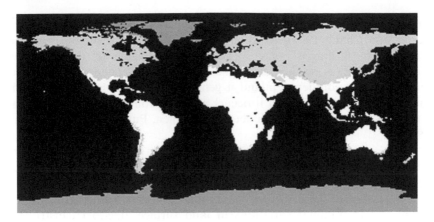

FIGURE 1.2 Approximate global distribution of snow cover, on a 1° grid. The lighter gray shows areas where snow cover sometimes occurs, the darker shade shows where the cover is essentially permanent. (Data for the northern hemisphere were calculated from data contributed by Richard L. Armstrong, National Snow and Ice Data Center, to the International Satellite Land-Surface Climatology Project (ISLSCP) Initiative II Data Archive (http://islscp2.sesda.com/ISLSCP2_1/html_pages/islscp2_home.html). Southern-hemisphere data were added using data from the MODIS snow and sea-ice global mapping project (http://modis-snow-ice.gsfc.nasa.gov/intro.html).)

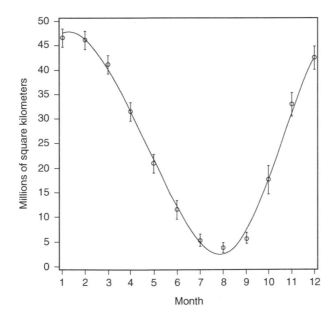

FIGURE 1.3 Average northern hemisphere snow cover, excluding Greenland, 1972–1991 (Robinson 1993). Circles show mean values; error bars show ±1 standard deviation in interannual variation. The curve is the best-fitting harmonic with terms at one and two cycles per year.

nonpermanent snow in the northern hemisphere is approximately 3×10^{15} kg (Foster and Chang 1993), or about $65 \, \mathrm{kg \, m^{-2}}$ over the snow-covered area, equivalent to 65 mm of water. The maximum value of this *snow water equivalent* (SWE) is reached at a time of year that depends on the location. For example, in southern Finland it generally occurs in February to March, while in the north of Finland it occurs about 2 months later, reaching values of typically a few hundred millimeters (Koskinen, Pulliainen, and Hallikainen 1997).

Snow cover is significant at all spatial scales. It represents an important geophysical variable for climate, especially through its role in controlling the Earth's albedo (Nolin and Stroeve 1997), and for hydrology (Ross and Walsh 1986; Barnett et al. 1989). The seasonal snow cover is responsible for the largest annual and interannual variations in land surface albedo (Armstrong and Brodzik 2002), and contributes an important feedback mechanism to the global climate system. Another important climatological effect is the thermal insulation provided by snow cover, which reduces the exchange of heat between the ground and the atmosphere. At a more local scale, snow cover is important from its potential for water storage for drinking, irrigation, and hydroelectric power generation, and for the production of floodwater (Rango 1993; Rango and Shalaby 1998). The thermal insulation provided by snow protects plants from low winter temperatures. Snow cover also provides economic benefit through winter recreation (e.g., skiing), as well as adverse economic impact through its potential to disrupt road and rail transport.

Thus there is a need to monitor the spatial extent of snow cover, as well as its depth and water equivalent. Measurement of the surface albedo is also important in modeling the energy balance of the snow pack.

1.3   ICE SHEETS AND GLACIERS

Permanent snow cover eventually forms a glacier, defined as an accumulation of ice and snow that moves under its own weight. The processes by which an accumulation of snow is transformed into glacier ice are discussed in Chapter 4. Glaciers occur in a huge range of sizes, with a ratio of approximately 10^8 between the areas of the largest and the smallest. The smallest glaciers are of the order of 10 ha ($10^5 \, \mathrm{m^2}$) in area, while the largest ice masses on Earth are the *ice sheets* of Antarctica and Greenland (Figures 1.4 to 1.6), representing between them about 99% of the mass and 97% of the area of land ice. Table 1.1 summarizes the global distribution of land ice. The total of $3.30 \times 10^{16} \, \mathrm{m^3}$ represents about 77% of the Earth's fresh water, the remaining 23% being groundwater (22%) and lakes, rivers, snow, soil moisture, and water vapor (1%) (Thomas 1993). The total discharge rate from the Earth's land ice has been estimated as $3 \times 10^{12} \, \mathrm{m^3}$ per year (Kotlyakov 1970) giving a mean residence time (the time for which a molecule of water remains as part of the glacier) of the order of 10^4 years (Hall and Martinec 1985). This figure can be

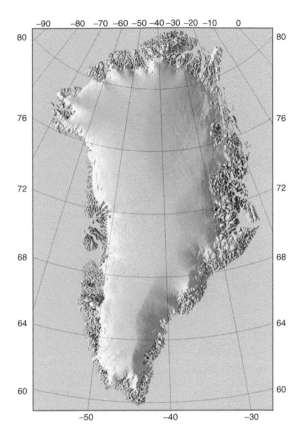

FIGURE 1.4 Shaded relief map of the Greenland ice sheet, provided by the Danish Polar Center (http://www.kms.dk/research/geodesy/index_en.html?nf=http://www.kms.dk/research/geodesy/geoid/geoid_en.html.)

contrasted with the residence times of weeks or months for temporary and seasonal snow cover.

As indicated by Table 1.1, the two ice sheets are enormously thick — on average over 1 km, and reaching a maximum thickness of 4500 m in the case of Antarctica, about 3000 m in Greenland. In fact, the Antarctic ice sheet is really two ice sheets, the East Antarctic and West Antarctic ice sheets, separated by the Transantarctic Mountains. The East Antarctic ice sheet is larger, thicker, and older than the West Antarctic ice sheet; the former is almost entirely above sea level, while the latter is almost entirely below it.

Like all glaciers, ice sheets flow in response to gravitational forces. Faster-moving regions of ice are called *ice streams*, essentially rivers of ice flowing across the slower-moving ice sheet. Ice also flows from the ice sheets in the form of *outlet glaciers* (Figure 1.7). When these reach the coast the ice continues to flow over the sea, forming an *ice shelf* that is anchored to the coast

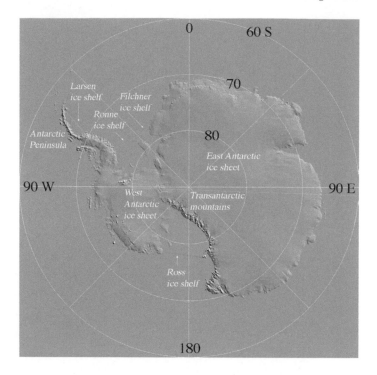

FIGURE 1.5 Shaded relief map of Antarctica, showing some major features (not all ice shelves are identified). (The map has been derived from the GTOPO30 global digital elevation model (http://edcdaac.usgs.gov/gtopo30/gtopo30.asp).)

FIGURE 1.6 Eastern margin of the Greenland ice sheet. In this area, mountain peaks emerge through the ice sheet as *nunataks*. (Photograph by courtesy of Richard Waller, School of Earth Sciences and Geography, Keele University.)

TABLE 1.1
Global Distribution of Land Ice

	Volume (m³)	% Total Volume	Area (m²)	% Total Area	Mean Thickness (m)	Sea Level Potential (m)
Antarctica	3.01×10^{16}	91	1.36×10^{13}	86	2.2×10^3	73.4
Greenland	2.60×10^{15}	8	1.73×10^{12}	11	1.5×10^3	6.5
Remainder	2.4×10^{14}	1	5×10^{11}	3	5×10^2	0.6
Total	3.30×10^{16}	100	1.58×10^{13}	100	2.1×10^3	80.5

Based on data given by Swithinbank (1985).

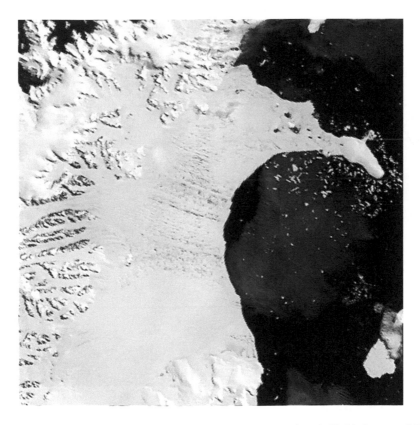

FIGURE 1.7 MODIS satellite image of part of the Larsen ice shelf, 31 January 2002. Several outlet glaciers can be seen feeding the ice shelf from the left. The central part of the ice shelf visible in this image was in the process of disintegrating when the image was acquired. (Image supplied by courtesy of Ted Scambos, National Snow and Ice Data Center, and downloaded from the NASA Earth Observatory web site at http://earthobservatory.nasa.gov.)

(Figure 1.7). Ice shelves are primarily an Antarctic phenomenon. Virtually all of the land area of Antarctica is covered by the ice sheet, while only about 80% of the area of Greenland is ice-covered. About 50% of the Antarctic coastline has ice shelves attached to it, the largest being the Ross ice shelf, between 200 and 1000 m thick and roughly the area of France. About 11% of the area and 2.5% of the volume of the Antarctic ice mass is represented by the ice shelves (Drewry 1983). Ice shelves undergo flexure as a result of ocean tides, and the seaward parts are eventually broken off to form *tabular icebergs*, which float away and ultimately melt. This process is normally in equilibrium with the addition of mass to the ice sheet in the form of precipitation, although there is now considerable evidence that the process is out of equilibrium in a number of cases, with ice shelves undergoing large-scale collapse.

Somewhat smaller* than the ice sheets are the *ice caps*, dome-shaped glaciers covering high-latitude islands or highland areas, and typically a few thousand square kilometers in extent (Figure 1.8). The major ice caps occur on

FIGURE 1.8 Landsat satellite image mosaic of Devon Island ice cap, Canada. (Image reproduced with permission of Global Land Ice Measurements from Space (GLIMS) Canadian Regional Center, University of Alberta.)

---

*Formally, an ice cap is up to 500,000 km^2 (5×10^{11} m^2) in area. Beyond this size it would be classed as an ice sheet.

FIGURE 1.9 Midre Lovénbreen, a valley glacier in Svalbard (author's photograph).

FIGURE 1.10 Marginal features of a piedmont glacier, Bylot Island, Nunavut, Canada. (Photograph taken by Ron DiLabio. Reproduced with permission of the Minister of Public Works and Government Services Canada, 2004, and by courtesy of Natural Resources Canada, Geological Survey of Canada.)

Iceland, in the Canadian Archipelago, Svalbard, and the islands of the Russian Arctic. The remaining classes of glacier are the *valley glaciers*, which form in mountain valleys, and *piedmont glaciers* which form when the ice from a glacier spreads out over flatter ground (Figures 1.9 and 1.10). Glaciers occur on all continents except Australia. The principal glaciated areas are Alaska, Iceland, Svalbard, Norway, the Russian Arctic islands, the Alps, southern Andes, Karakoram, and Himalaya mountains (Sugden and John 1976). Of the Earth's glaciated area, excluding the Antarctic and Greenland ice sheets, only a few percent occurs in the southern hemisphere, predominantly in Chile (Rivera et al. 2002).

As with snow cover, glaciated regions are significant at a wide range of spatial scales. The range of relevant temporal scales is also broad. The large ice sheets are immensely important in the global climate system, occupying about 11% of the Earth's land surface. Their very high albedo (which can exceed 90%) makes them brightest naturally occurring objects on Earth (Bindschadler 1998) and hence an important component of the albedo feedback mechanism of the global climate system. Their low temperature contributes to the global temperature gradient that drives the atmospheric circulation system, and their size and position profoundly affect the details of this circulation (Bindschadler 1998). They are major repositories of fresh water with, as indicated in Table 1.1, the potential if completely melted to raise the global sea level by about 80 m. The importance of the balance between frozen and liquid water on a global scale is illustrated by considering past variations. During the most recent glacial maximum, about 18,000 years ago, the total volume of ice was about 2.5 times, and the total area about three times, their present values (Flint 1971), with the consequence that sea level was about 125 m below its present level (Fairbanks 1989). During the last glacial minimum, about 120,000 years ago, sea level was about 6 m higher than today, and it is possible that the Greenland ice sheet largely disappeared (Koerner 1989). Thus, the most obvious manifestation to be expected from land ice in response to global climate change is a change in ice volume leading to an alteration of sea level. This has been rising by 1 to 2 mm per year since 1900, and although not all sources of this water have been identified (Meier 1990) it seems probable that up to half of it can be attributed to the melting of the smaller Alpine glaciers (Meier 1984). Knowledge of the contribution from the major ice sheets remains poor, though it is likely that the Greenland ice sheet will contribute more than Antarctica, since the former experiences widespread summer melting while the latter does not (Bindschadler, Fahnestock, and Kwok 1992). Thus, monitoring of the phenomenon of surface melting, and the characteristic surface zonation that this produces* (discussed in Chapters 4 and 8), is highly desirable, especially for the smaller and more dynamic valley glaciers, the Greenland ice sheet, and the Antarctic Peninsula region, which is located in a frontal zone and so characterized by large climatic gradients (Rau and Braun 2002).

Other large-scale phenomena that may be associated with global climate change include the possible destabilization and disintegration of the Antarctic ice shelves. In particular, it has been suggested that the West Antarctic ice sheet may be unstable (Oppenheimer 1998). Dramatic *calving* events (breaking away of ice) have been monitored using satellite data. The Larsen ice shelf has been retreating since the 1940s, increasingly rapidly since about 1975. Major calvings have occurred since 1986 (Rott et al. 2002), associated with

---

*Even quite small temperature changes could be expected to produce large lateral shifts in the position of zone boundaries because of the small surface slopes that characterize ice sheets.

acceleration of the glaciers that formerly fed the ice shelf, and the retreat is probably now irreversible (Doake et al. 1998; Scambos et al. 2000).

As well as their role as indicators of climate, smaller glaciers are also economically important as sources of fresh water and for hydroelectric power generation in Norway, Iceland, parts of the United States, and the Alps (Hall and Martinec 1985; Brown, Kirkbride, and Vaughan 1999). They can represent hazards through a variety of mechanisms (Williams and Hall 1993) including advance, retreat and surges,* jökulhlaups (release of ice-dammed lakes), iceberg discharge, and rapid disintegration.

There is thus a need to measure and monitor a range of properties of ice sheets and glaciers, including the ice volume and extent, distribution of surface features related to the temperature and wind regime (Bromwich, Parish, and Zorman 1990; Frezzotti et al. 2002), dynamics, and mass balance. As with snow cover, measurement of the surface albedo is also important, since it gives the possibility of modeling the energy balance of a glacier. Spaceborne techniques offer major advantages over *in situ* measurements in all of these cases.

1.4 ICEBERGS

Icebergs are masses of freshwater ice that have calved from a glacier or ice shelf and fallen into the sea or a body of fresh water, or that have been produced as a result of the breaking up of larger icebergs. Evidently, icebergs are produced only from ice masses that terminate in water. Icebergs are classified according to both size and shape, as described in Chapter 4, and can range in size from fragments of a meter or so to tens or even hundreds of kilometers. The largest are the *tabular icebergs* (Figure 1.11) that calve from ice shelves. These are more common in the Antarctic than the Arctic.

The rate at which icebergs are produced is highly variable, since it is influenced by glacier velocity, including surge and recession events, and the degree of crevassing (Løset and Carstens 1993; Jensen and Løset 1989), ocean conditions (Vinje 1989; Dowdeswell et al. 1994), and the extent of sea ice. Most North Atlantic icebergs originate from the western side of the Greenland ice sheet, which calves about 10,000 icebergs a year; a few are also contributed from the eastern Canadian archipelago. Other sources of icebergs in the northern hemisphere include the Franz Josef Land archipelago, Severnaya Zemlya and Svalbard, and Alaska. In the southern hemisphere, most icebergs originate from the Antarctic ice sheet. The total number of icebergs in the Southern Ocean has been estimated at 200,000 (Orheim 1988), representing

---

*Glacier surging is a rare phenomenon, displayed by about 1% of all glaciers (Jiskoot, Boyle, and Murray 1998). It is particularly common in Svalbard. During a surge, the flow velocity increases by a factor of 10 to 1000 for a short period (Murray et al. 2002), reaching values of 100 m/day or more.

FIGURE 1.11 A tabular iceberg (http://www.genex2.dri.edu/gallery/ice/tabular3.htm.)

about 93% of the global iceberg mass, of the order of 10^{15} kg (and hence comparable to the total mass of nonpermanent snow).

About 15 to 20% of the volume of a newly calved iceberg is above the waterline, because the upper part of the iceberg is less dense than the bulk. However, once an iceberg has begun to weather it is likely to roll over, causing the lower-density parts to melt. In this case, the volume above the waterline is around 8%. The lifetime of an iceberg is strongly dependent on the ocean drift that carries it into warmer waters; 3 years is typical. The frequency of occurrence of icebergs decreases with distance from the point of production. For example, Ebbesmeyer, Okubo, and Helset (1980) describe an almost linear variation of iceberg frequency with latitude on a transect from Baffin Bay to the Grand Banks, falling to zero at around 46° north.

Most of the ice lost from the Antarctic ice sheet is in the form of iceberg calving (Jacobs et al. 1992). The study of iceberg calving rates thus plays an important role in understanding the mass balance of the ice sheet, and the West Antarctic ice sheet is the subject of particular scrutiny because of its recent production of numbers of giant icebergs (Lazzara et al. 1999). Monitoring of position of the edge of an ice shelf gives information on its rate of growth, and hence the iceberg discharge cycle (Fricker et al. 2002). The motion of a floating iceberg is governed by ocean currents, tides, wind,* and bathymetry. Icebergs with deeper drafts are more strongly influenced by ocean currents, while

---

*Wind forces have an appreciable effect on icebergs only when the wind speed exceeds about $10\,\mathrm{ms}^{-1}$ (Løset and Carstens 1993).

smaller bergs are sensitive to wind forces and wind-driven surface currents (Gustajtis 1979). The rate of disintegration of an iceberg is governed by water temperature, sea state, and direct solar radiation as well as the iceberg's size and shape.

Large icebergs can also be used as floating research platforms for manned investigations, or equipped with instruments that return data by telemetry. A typical arrangement includes a water temperature sensor and perhaps also an atmospheric pressure sensor, and uses the ARGOS service to transmit its data and determine its position (Løset and Carstens 1993; Murphy and Wright 1991; Tchernia 1974; Tchernia and Jeanin 1982) (see Figure 1.12). This approach has been adopted by the International Ice Patrol (IIP) for tracking icebergs since 1976. More recently, the Global Positioning System (GPS) has been used for determination of position, since it gives considerably greater precision than the ARGOS system. Tracking of buoys on drifting icebergs provides the possibility of mapping the field of ocean circulation. However, it would be impractical to deploy enough of these drifting buoys to provide adequate spatial resolution of ocean circulation patterns. For example, it has been estimated (FENCO 1987) that a mean density of 400 buoys per $250 \times 250 \, km$ area per year would be needed to provide adequate resolution of an eddy field. A lower estimate (less by a factor of 3) has been proposed by Venkatesh, Sanderson, and El-Tahan (1990), but even so, tracking by satellite remote sensing also provides an important approach to the problem.

Icebergs notoriously represent a hazard to shipping, fishing, offshore hydrocarbon extraction etc. Fixed oil and gas extraction facilities are particularly inconvenienced by icebergs (Rossiter et al. 1995), since the cost of uncoupling a rig and towing it out of harm's way is very high. However, all forms of maritime operation are potentially jeopardized by the presence of floating ice, and a number of monitoring programs have been established

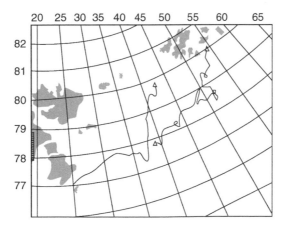

FIGURE 1.12 Two iceberg drift tracks from Franz Josef Land to Svalbard, recorded using the ARGOS tracking system. (Reproduced from Spring, Vinje, and Jensen 1993.)

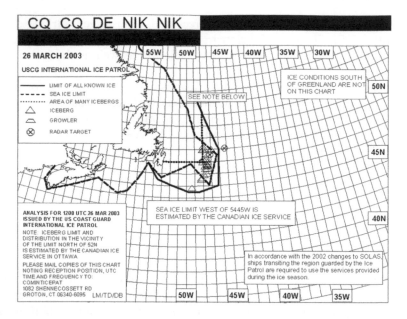

FIGURE 1.13 Typical ice chart produced by the International Ice Patrol, showing the distribution of sea ice and icebergs. Although the chart is a typical product of the IIP, the sea ice conditions were extreme. The extent of the ice off Nova Scotia and at the entrance to the Gulf of St. Lawrence on this date was the most severe for two decades. (Image and interpretation reproduced by courtesy of the U.S. Coast Guard International Ice Patrol.)

as a result, including the IIP, which was set up in response to the sinking of the *Titanic* (Figure 1.13). Similarly, a group of oil company operators formed the organization "Operator Committee North of 62°" (OKN) in the late 1970s to obtain environmental data from the Norwegian offshore region. In 1987, OKN inaugurated the Ice Data Acquisition Program (IDAP) to collect data on sea ice and icebergs in the Barents Sea.

Until about 1990, most large icebergs in the Arctic were located by visual observation during aerial reconnaissance (Jeffries and Sackinger 1990). As a consequence, the detection and tracking of large icebergs has not been systematic (Jeffries and Sackinger 1990). Since about 1990, the use of remotely sensed data has played an increasingly significant part in the strategic and tactical monitoring of icebergs.

Icebergs have also been proposed as a freshwater resource, by towing them to areas that need water (Holden 1977).

1.5 SEA ICE

Sea ice is formed when the temperature of seawater is reduced below about $-1.8°C$ (this figure depends on the salinity of the sea water). It is a significant

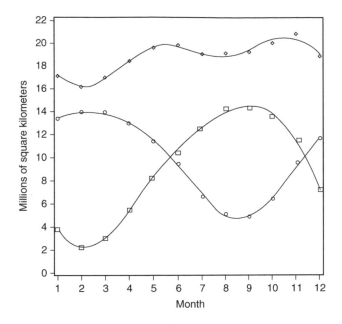

FIGURE 1.14 Estimated sea ice cover for 1995 in the northern hemisphere (circles), southern hemisphere (squares) and globally (lozenges). The curves are best-fitting harmonic series with terms up to two cycles per year. (Calculated from data contributed by Richard L. Armstrong, National Snow and Ice Data Center, to the International Satellite Land-Surface Climatology Project (ISLSCP) Initiative II Data Archive (http://islscp2.sesda.com/ISLSCP2_1/html_pages/islscp2_home.html).)

phenomenon in both the northern and southern hemispheres, although the very different geographical settings of the Arctic and the Antarctic give it different characteristics.

The sea ice cover in the Southern Ocean varies from about 18 million km^2 in September to less than 4 million km^2 in February. In the Arctic the cover varies from about 8 million km^2 in September to 15 million km^2 in March (see Figure 1.14). These figures are, however, rather variable from year to year. Up to about 0.5 million km^2 in the Arctic, and 1 million km^2 in the Antarctic, can consist of ice less than about 0.3 m in thickness (Grenfell et al. 1992). However, the mean thickness is a few meters, which implies that the total mass of sea ice is of the order of 3×10^{16} kg, about one order of magnitude greater than the global mass of nonpermanent snow and three orders of magnitude less than that of land ice.

Figure 1.15 shows the typical maximum and minimum distributions of sea ice. Permanent sea ice cover in the northern hemisphere is confined roughly to latitudes north of 82° in the Eurasian sector of the Arctic, and 75° in the North American sector, while in the Antarctic it is largely confined to the Weddell Sea. The maximum extent of sea ice is limited approximately by latitudes 60° north and south. In the Southern Ocean this limit coincides more or less

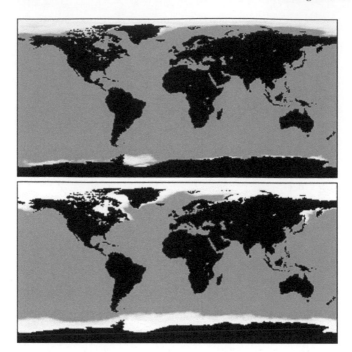

FIGURE 1.15 Minimum (top) and maximum (bottom) concentrations of sea ice observed in 1995. The concentrations are represented as shades of gray, with white representing 100%. (Data source as for Figure 1.14.)

with the Antarctic Convergence. In the northern hemisphere, the warm North Atlantic Drift current reduces the extent of the sea ice in the Norwegian and Barents Seas, while ice cover extends as far south as 45° on the eastern American coast.

Sea ice can be broadly classified into new ice, first-year ice, and multiyear ice. *New ice* is a term embracing the various stages through which ice is transformed from individual ice crystals to consolidated ice floes that are of the order of a meter thick. *First-year ice* (Figure 1.16) is ice that has not yet been exposed to the summer melt period, while *multiyear ice* has survived at least one summer (Figure 1.17). (The term *second-year ice* is sometimes also used.) As the ice ages, it tends to become thicker and rougher as a result of plastic deformation when the floes collide with one another under the influence of wind and ocean currents. As indicated by Figure 1.14, multiyear ice is more widespread in the Arctic than the Antarctic, where it is largely confined to the Weddell Sea.

Sea ice plays an extremely important role in the global climate system (Carsey, Barry, and Weeks 1992; Holt, Rothrock, and Kwok 1992), and has been integrated into climate models since about 1990. It has a high albedo and so contributes a major feedback to the Earth's radiation budget. As a largely impermeable layer between the atmosphere and the ocean, it modifies the

FIGURE 1.16 First-year sea ice in the Ross Sea, with penguin in foreground. (Photograph by Michael van Woert, NOAA (http://www.photolib.noaa.gov/corps/corp2632.htm.)

transport of heat, water vapor, and other gases between these two reservoirs. As a physically massive object, it modifies the processes by which energy and momentum are exchanged between the atmosphere and the ocean. It also has a profound effect on the global system of ocean circulation. As the ice ages, it rejects cold salty water into the ocean, which sinks and then flows along the ocean floor. This "thermohaline convection" is the main driver of the global

FIGURE 1.17 Multiyear sea ice in the Beaufort Sea, showing a pressure ridge between two floes. (Photograph by courtesy of J. Dana Hrubes, South Pole Station, Antarctica.)

"conveyor belt" of ocean circulation and is largely responsible for the surface flow of warmer water into the North Atlantic.

Sea ice also has biological significance. It provides a habitat for microorganisms and a platform for sea mammals and penguins (and the occasional polar expedition), and it reduces light penetration into the ocean (Parkinson 2002). Similarly to icebergs, it also represents a hazard to marine transport and to the operation of fixed facilities such as oil rigs, especially in the *marginal ice zone* where the ice is at its most dynamic in response to forcing by wind and water currents. Conversely, during the Cold War era, sea ice areas provided a strategic hiding place for nuclear submarines. When deeply submerged these are only detectable acoustically, so a noisy environment (and the ocean beneath an ice cover is noisy as a result of the distortion of floes and the collisions between them) is a favorable place for concealment.

As well as being a major part of the global climate system, sea ice also functions as an indicator of climate. Since it has a survival time (the typical time between formation and melt) of the order of 1 year, the ice cover effectively integrates the climate signal over this period, in contrast to the ice sheets, which have much longer integration times. The cover is comparatively widespread, and occurs in the regions where models predict the greatest amplitude of climate change. Sea ice is thus expected to show early indications of such changes (Budyko 1966; Alley 1995; Stouffer, Manabe, and Bryan 1989), and in fact the ice cover has been observed to be retreating by about 0.3% per annum over the last few decades (Bjørgo, Johannessen, and Miles 1997; Parkinson et al. 1999), while deep-water sea ice may have thinned by the order of a meter in 40 years (Rothrock, Yu, and Maykut 1999; Wadhams and Davis 2000). It has been proposed that a possible consequence of global climate change is that the ocean circulation conveyor belt could in effect be switched off.

The principal variables needed to describe the sea ice cover are (Carsey, Barry, and Weeks 1992) its extent, thickness, motion, the distribution of floe sizes and of the *leads* (areas of open water) between them, as well as characterization of the snow cover and the presence of pools of meltwater on the surface of the floes. These are all susceptible to measurement by remote sensing methods, although ice thickness cannot currently be measured by spaceborne observations.

1.6 FRESHWATER ICE

Lakes and rivers at high latitudes freeze during the winter. As with snow cover, this is primarily a northern hemisphere phenomenon. In Asia, freezing usually occurs at about the beginning of November at latitudes 50 to 60°, and about a month earlier 10° further north. The ice breaks up around the beginning of May at latitudes 50 to 60°, and about a month later 10° further north. In North America the spatial distribution is rather more complicated, since there are strong east–west gradients in the dates of freezing and thawing, as well as the latitudinal gradient (Figure 1.18). The longest duration of freezing is about 8 months, occurring in the northern Taimyr Peninsula in Siberia.

The formation and thawing of river and lake ice (Figures 1.19 and 1.20) has considerable hydrological significance. The flow of a river, and hence the freshwater input to the lake or sea fed by it, is reduced or even stopped altogether during the frozen period, and flow rates during the thaw can be extremely high. If a river is blocked by ice floes or by a build-up of superimposed *aufeis* during the thaw period, water can flood onto the surrounding land.

Freshwater ice provides a valuable climatic indicator (Palecki and Barry 1986; Maslanik and Barry 1987; Wynne and Lillesand 1993). Ground-based

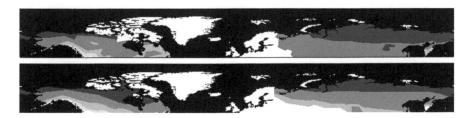

FIGURE 1.18 Dates of river freezing and thawing of rivers and lakes in the northern hemisphere (based on Mackay and Løken 1974). (Above) Autumn freezing. The progressively darker shades of gray indicate January, December, November, and October. (Below) Spring thawing. The progressively darker shades of gray indicate March, April, May, and June. White areas indicate no data.

FIGURE 1.19 (See Color Figure 1.2 following page 108) River ice. (Photograph by courtesy of Professor F. Hicks).

FIGURE 1.20 Ice on Toolik Lake, Alaska, 15 June 2003. (Reproduced by courtesy of Richard Flanders, Toolik Field Station, Alaska.)

observations of the dates of freezing and break-up have shown that these are well correlated with air temperature during these seasonal transitions. Freezing and thawing dates change by approximately 4 to 7 days for each degree (Celsius) change in air temperature.

The formation of river and lake ice has both positive and negative economic consequences. While it can provide winter routes for land transport, it also impedes water transport and can damage harbor and shore structures and moored boats. Fish and other freshwater species are also affected. A lake that regularly freezes to the bottom is not viable for fish, and a fish population in a lake or river that does not normally freeze completely will be killed if such freezing does occur.

The most important freshwater ice variables to be studied, then, are the extent of ice cover, its thickness, and whether it is frozen to the bottom. The task is made more difficult by the facts that freezing and thawing can be rapid events requiring high temporal resolution to represent them with sufficient precision, and that most rivers and lakes will require high spatial resolution from an observing system.

1.7 PERMAFROST

The last component of the cryosphere to be considered is permafrost. This is ground that remains frozen throughout the year.* The variation of ground temperature with depth is controlled by two main factors. The first of these is the geothermal gradient, as a result of which the temperature increases with depth. The second is the annual fluctuations in the surface temperature, which cause a seasonal variation in ground temperature, attenuated with depth, that is superimposed on the geothermal gradient. If the maximum surface temperature is low enough, these factors can combine in such a way that the ground temperature never exceeds 0°C over some range of depths. This is the permafrost layer. Above the permafrost is the *active layer*, which experiences summer thawing. In regions where permafrost occurs, the active layer is typically a few meters deep, while the permafrost layer below it is typically tens of meters in depth, although depths of several hundred meters, and even over 1000 m, have been reported in some locations (Washburn 1980). The phenomenon of permafrost is controlled largely by climate, although slope and aspect of the terrain, soil type, drainage, vegetation, and snow cover are also significant.

Permafrost is largely a northern hemisphere phenomenon — about a quarter of the northern hemisphere landmass is underlain by it — although it also occurs in mountainous regions throughout the world (Williams and Smith 1989; Zhang et al. 1999). Four classes of permafrost are conventionally distinguished, based on its spatial distribution: continuous, discontinuous, sporadic, and isolated permafrost. It is also useful to distinguish the proportion of ice contained within it. Figure 1.21 shows a simplified northern hemisphere permafrost map.

The distribution of permafrost can be altered by mechanical or thermal disturbance of the soil, and this sensitivity makes its presence a major factor to

---

*Formally, permafrost is defined as material that remains below 0°C for at least 2 years.

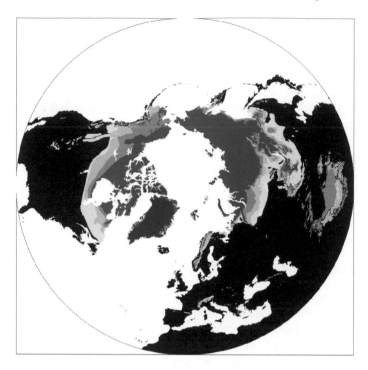

FIGURE 1.21 Distribution of permafrost in the northern hemisphere (derived from Brown et al. 1998). Progressively darker shades of gray represent relic, isolated, sporadic, discontinuous, and continuous permafrost. The darkest gray represents ice.

be considered in the construction of buildings, roads, and other infrastructure at high latitudes. Many buildings have collapsed as a result of the partial melting of the permafrost on which they stood.

Like all other components of the cryosphere, permafrost is a significant indicator of the global climate system, and the study of the extent of the various permafrost zones and the depth of the active layer is an important aspect of climate monitoring activities. The greater part of the world's permafrost is only a few degrees below freezing point, so that it has the potential to be a sensitive indicator of temperature changes (Beltrami and Taylor 1994; Lachenbruch and Marshall 1986). Permafrost has been generally warming and retreating northward since the end of the Little Ice Age (Laberge and Payette 1995; Wang et al. 2000), although with considerable spatial and temporal variability (Serreze et al. 2000).

Although the presence of permafrost is indicated by surface features such as polygonized terrain and thermokarst, which can be studied by remote sensing methods, the cryospheric component itself is buried beneath the surface and not amenable to remote sensing in the sense in which it is defined in this book. Remote sensing of permafrost will therefore not be considered.

2 Remote Sensing Systems for Observation of the Earth's Surface

2.1 INTRODUCTION

In general terms, remote sensing can be interpreted as the gathering of information about an object without physical contact. In the more useful but more restricted sense in which the term is normally employed, it refers to airborne or spaceborne observations using electromagnetic radiation. This radiation is either naturally occurring, in which case the system is said to employ *passive* remote sensing, or is generated by the remote sensing instrument itself (*active* remote sensing).

Naturally occurring radiation includes reflected solar radiation, which is largely confined to the visible and near-infrared parts of the electromagnetic spectrum (wavelengths between roughly 0.35 and 2.5 μm) and thermally emitted radiation. The range of wavelengths generated by a thermally emitting body depends on the temperature. The dominant wavelength is approximately A/T, where T is the absolute temperature of the body and A is a constant with a value of about 0.003 Km, so for a body at a temperature of 273 K (0°C) the dominant wavelength is around 11 μm. The distribution of energy over wavelengths falls very sharply for shorter wavelengths, but at longer wavelengths the decrease is much more gradual. For this reason, thermally emitted radiation can be detected both in the *thermal infrared* part of the electromagnetic spectrum (wavelengths typically 8 to 14 μm) and also in the *microwave* region (wavelengths typically 1 cm to 1 m). The wavelength region between 14 μm and 1 cm is largely blocked by the atmosphere. Figure 2.1 summarizes the useful regions of the electromagnetic spectrum for remote sensing and the transparency of the atmosphere.

Passive remote sensing systems that detect reflected solar radiation are designed to measure the *radiance*, i.e., the amount of radiation reaching the sensor in a particular waveband. If the amount of radiation that is incident on the Earth's surface is known, the *reflectance* of the surface can be calculated (this requires that the effects of the atmosphere should be corrected). This can be said to be the primary variable that is measured by such systems. The goal of subsequent analysis of the remotely sensed data is to interpret the value of this reflectance and its spatial and temporal variations.

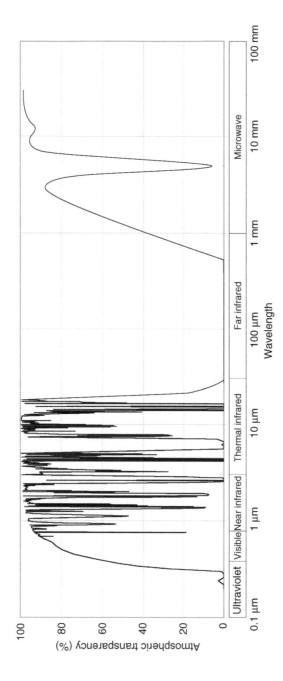

FIGURE 2.1 Typical atmospheric transparency and principal regions of the electromagnetic spectrum. These regions can be further subdivided. The visible band is conventionally divided into the spectral colors from violet to red, while the microwave region is often designated by the names of radar bands: P (centered around 1 m), L (400 mm), S (100 mm), C (60 mm), X (40 mm), K_u (18 mm), K_l (14 mm), K_a (9 mm).

In the case of remote sensing systems that detect thermal radiation, the sensor again measures the radiance reaching the instrument in a particular waveband. This radiance is normally expressed as a *brightness temperature*, which is the temperature of a perfect emitter (a so-called *black body*) that would produce the same amount of radiation. The primary variable to be determined is the brightness temperature of the radiation leaving the Earth's surface (again, the effects of the atmosphere must be taken into account to relate the brightness temperature measured at the sensor to the brightness temperature leaving the surface). The brightness temperature is related to the physical temperature of the surface and its *emissivity*, a unitless quantity that defines the ratio of the actual radiance to the radiance that would be emitted by a black body at the same temperature. It follows that the emissivity must therefore take a value between zero and unity, and that it is unity for a black body. Clearly, if the emissivity of a material is known and its brightness temperature can be measured, its actual physical temperature can be calculated.

Active remote sensing systems can be divided into two main types. Firstly, there are *ranging* instruments, whose primary purpose is to measure the distance from the sensor to the Earth's surface by measuring the time for short pulses of radiation to travel down to the surface and back again. From this information, the Earth's surface topography can be investigated. The radiation is visible light or (more usually) near-infrared in the case of laser profilers, while radio-frequency radiation is used by radar altimeters, impulse radar, and similar systems. The second type of active system is, similarly to the passive reflective systems, primarily designed to measure the surface reflectance. However, instead of relying on incident solar radiation, the instrument illuminates the Earth's surface and analyzes the signal returned to it. This gives the possibility of much greater flexibility in the characteristics of the radiation — its direction, wavelength, polarization, and time-structure (for example, it can be pulsed) can all be controlled. Systems that employ this approach can generally be classed as *imaging radars*, since they operate in the microwave part of the electromagnetic spectrum. The primary variable measured in this case is the *backscattering coefficient*, a unitless quantity related to the concept of reflectance and usually specified in decibels (dB) rather than as a simple ratio or percentage. Its dependence on the imaging geometry is often important.

In this chapter we will examine the principal types of remote sensing system in greater detail, presenting a theoretical outline and giving examples of important real systems. This treatment will necessarily be fairly brief. A fuller treatment can be found in general works on remote sensing (e.g., Kramer 1996; Campbell 1996; Rees 1999, 2001). The ordering of the material in this chapter is in general from the more to the less familiar, from passive to active, and in the direction of increasing wavelength from the visible region through the infrared to the radio wavelengths used in radar systems. We begin with a discussion of aerial photography, the historical prototype of remote sensing and still significant in its own right. (More information than can be presented here will be found in, e.g., Avery and Berlin 1992; Campbell 1996.)

2.2 AERIAL PHOTOGRAPHY

Aerial photography can be characterized as a passive imaging technique operating in the visible and reflective (near-) infrared parts of the electro-magnetic spectrum.

2.2.1 PHOTOGRAPHIC FILM

Photography is unique among remote sensing systems in that the mecha-nism for detecting electromagnetic radiation is a photochemical one.[*] The photographic film constitutes the array of detectors. This consists, in its simplest form, of a suspension of tiny crystals (usually referred to as *grains*, with the suspension being referred to as an *emulsion*) of silver halides (bromide or iodide) in a porous gelatin matrix, supported on a thin plastic film. Exposure to light can convert a few silver ions in a crystal into metallic silver; this is termed a *development center*, and is thermodynamically stable against reionization. The exposed film is said to contain a *latent image*. Chemical processing (*development*) of the exposed film converts entire crystals to metallic silver if a development center is present, but not otherwise, and then removes the remaining crystals. The result of this is that the opacity of the exposed and developed film depends in a characteristic way on the amount of light to which it was exposed. Since the film is more opaque in regions that were more heavily exposed to light, it is termed a *negative*.

This very brief description has omitted a number of important points. The film as described would respond only to ultraviolet radiation, since only photons of ultraviolet light would be sufficiently energetic to produce the necessary development centers. In practice, the emulsion layer contains sensitizing dyes to extend the spectral response into the visible and sometimes also the near-infrared region. By the use of suitable filters the spectral response can be limited to, for example, just the near-infrared, and by constructing stacks of three emulsion layers with sensitivities to different spectral ranges, color film can be produced. The principal types of film in terms of spectral response are *panchromatic film*, which has a more or less uniform response across the visible region, *infrared film*, which responds only to the near-infrared region, *color film*, which is separately sensitive to red, green, and blue light, and *color infrared film*, which is separately sensitive to near-infrared, red, and green light. Color infrared film is normally printed as *false color infrared* (FCIR), in

---

[*]Digital photography is generally supplanting film photography, although at present the photochemical process is still preferred for aerial photography.

which the areas that have responded to infrared radiation are printed in red, those that have responded to red light are printed in green, and those that have responded to green light are printed in blue.

The performance of photographic film is usually described by specifying its *resolution, speed,* and *contrast.** All of these can be understood at a basic level in terms of the simple model of film response that was outlined above. The spatial resolution of the film is clearly determined by the size of the halide grains, such that finer grains give higher resolution. In the photographic literature the resolution is usually specified in terms of the number of *line-pairs* per millimeter that can be resolved on the negative, or more comprehensively by specifying the *modulation transfer function* of the film. However, we adopt a rather simpler definition by specifying the width of the *point spread function,* i.e., the diameter of the region on the negative produced by light focused to a spot from a single direction.† Typical values of this figure range between about 10 μm for a very high resolution film, capable of resolving 300 to 600 line-pairs per millimeter (such a film might be used for high-altitude aerial photography) and about 20 μm for a high-speed aerial photography film (perhaps 25 to 100 line-pairs per millimeter). For comparison, a consumer-grade 35 mm film might have a FWHH resolution of around 25 to 30 μm.

Film speed is a measure of the *exposure* (defined as the product of the illuminance at the film with the time of the exposure, and hence measured in lux-seconds) required to produce a significant degree of opacity in the developed film. A fast film requires a small exposure, and conversely. Since an entire grain can be converted to silver if a development center is present within it, a coarse-grained film will be faster than a fine-grained one. Thus the desirable characteristics of high speed and high spatial resolution are to some extent incompatible, and the optimum choice of film type will depend on the application.

The contrast of a film refers to the range of exposures to which it responds. If the exposure is too low, essentially no development centers will be generated and the developed film will be fully transparent, while if the exposure is high enough, all grains will be converted to silver on development, giving a fully opaque negative. A high-contrast film is one for which the ratio of the maximum to minimum useful exposure is small. This can be achieved by arranging that the grains are practically all the same size. Conversely, a low-contrast film, which responds to a wide ratio of exposures, has a broad distribution of grain sizes. The optimum contrast for a film will depend on the range of brightness present in the scene: a high-contrast film would be preferable for a scene with little variation in brightness.

---

*Technical specifications of aerial photography and other films can usually be found on manufacturers' web sites.

†Formally, we define the *full width to half height,* or FWHH, which is the diameter of the region in which the optical density is within half of its maximum value.

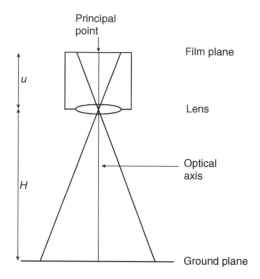

FIGURE 2.2 Schematic optics of a vertical aerial photograph.

2.2.2 LENS OPTICS

Figure 2.2 shows schematically the geometry of a vertical aerial photograph. The lens is at a distance H above the ground, and rays from the ground are brought to a focus in the film plane a distance u beyond the lens. H and u are related to the focal length f of the lens by the following formula:

$$\frac{1}{u} + \frac{1}{H} = \frac{1}{f} \qquad (2.1)$$

In practice, since $H \gg f$, it follows that $u \approx f$.

The figure shows three "construction rays" passing through the center of the lens, and hence undeviated. From a consideration of similar triangles it is thus clear that the linear *scale s* of the negative is given by

$$s = \frac{u}{H} \approx \frac{f}{H} \qquad (2.2)$$

The spatial resolution of the photograph in the ground plane can obviously be deduced from the scale and the spatial resolution in the negative. Similarly, the scale determines the spatial coverage of the photograph. For example, if the negative is 240 mm square (this is termed the film's *format*), the camera has a focal length of 150 mm and the height H is 3000 m, the scale s is 0.00005 (1:20,000) and the coverage is 240 mm × 20,000 = 4800 m. If the spatial resolution on the film is 20 μm, the corresponding resolution at the ground plane is 0.4 m.

We should make two more remarks about lens optics in photography. The first is to observe that Figure 2.2 is indeed very schematic, and suggests that focusing is achieved using a single lens, whereas a compound lens consisting of several elements is much more usual. Nevertheless, the concept of a single effective focal length remains valid. The second remark is that the spatial resolution of the camera is determined by both the film and the optics. The spatial resolution of the best-quality optical components is set by the diffraction limit, such that the angular FWHH is approximately λ/D, where λ is the wavelength of the radiation and D is the diameter of the lens. Thus the contribution to the FWHH at the film plane is approximately $f\lambda/D$. It is clear that the overall spatial resolution could be significantly degraded in cases where both the film resolution and the value of f/D are large.

2.2.3   GEOMETRIES OF AERIAL PHOTOGRAPHY

Most aerial photography that is used for quantitative analysis is obtained using a *metric camera* (or *cartographic camera*), which has very well defined imaging geometry and is operated with the optical axis vertical. The great advantage of vertical aerial photography is that since all points in the ground plane (assumed to be level) are at the same depth H below the camera, the scale in the ground plane is constant. Metric cameras are designed to minimize distortion, and in addition the relationship between an object's position on the ground and the corresponding position of the image on the negative is known, through calibration, with great accuracy. This is termed the *camera model*, and is necessary for accurate geometrical mapping from the photograph. *Fiducial marks* are recorded on the negative to allow the precise determination of coordinates from it.

Coverage of larger areas can be obtained in a number of ways. The simplest is *oblique* photography, in which the optical axis is not vertical. The principal disadvantage of this approach is that the scale in the ground plane is no longer constant (Figure 2.3). Other approaches to covering larger areas are also possible, particularly the use of panoramic cameras, strip cameras, and reconnaissance cameras. All of these are characterized by lower geometric accuracy and, in general, less precisely known camera models, than the metric cameras.

2.2.4   MAPPING, RELIEF, AND STEREO- AND ORTHOPHOTOS

One of the major advantages of aerial photography is the geometric fidelity* of the negative, especially in the case of a vertical photograph acquired with a

---

*This is in contrast to the radiometric properties of the negative. Apart from the fact that the opacity of the negative is a monotonic function of the exposure, the relationship between these two quantities can be influenced by the way in which the film is processed.

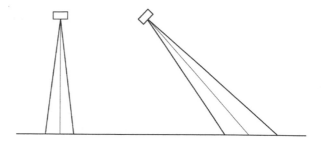

FIGURE 2.3 Oblique photography provides greater coverage at the expense of variable scale.

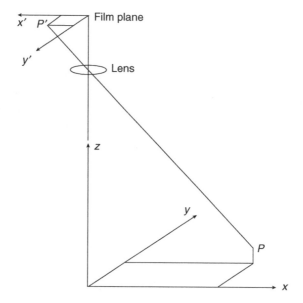

FIGURE 2.4 Geometry of relief displacement in a vertical photograph.

metric camera. If the ground surface is truly planar and perpendicular to the optical axis, it is clear that the coordinates of objects in this plane can be determined immediately from measurements on the negative once the camera model and scale are known. However, these conditions will often not be met, because of a misalignment of the camera's optical axis, the presence of significant relief in the surface being viewed, or both. The effect of these phenomena is illustrated in Figure 2.4.

In Figure 2.4, xy is the nominal ground plane and the z axis points toward the camera. The position of a point P in the scene can thus be expressed by its Cartesian coordinates (x, y, z). The coordinates of the corresponding point P' in the film plane are (x', y'). If the z-coordinate of the camera lens is H and the distance from the lens to the film plane is f (strictly u), simple geometric considerations show that

$$x' = \frac{fx}{H - z} \tag{2.3}$$

and

$$y' = \frac{fy}{H - z} \tag{2.4}$$

Thus we can see that the imaged point P' has been displaced from the position it would have if $z = 0$, unless it happens to coincide with the principal point. This phenomenon is known as *relief displacement*, and it must be corrected if the photograph is to be used as a map with a constant scale.

If the object in the scene is vertical, and both its top and its base (on the ground plane) are visible in the image, equations (2.3) and (2.4) can be used twice — once each for the top and the base — to deduce both the x and y coordinates of the object and its height z. While this is possible in urban areas where the object might be, for example, the vertical side of a building, it is unlikely to occur in the natural environment. In this case, we do not know the values of x and y because we cannot identify the point in the image corresponding to $z = 0$. This problem is solved by *stereophotography*, in which two vertical photographs are taken of the same scene from different positions. To see this formally, we suppose that the first photograph is taken with the geometry shown in Figure 2.4, and the second photograph is taken from the same height but with the camera lens above the point $(X, Y, 0)$. The coordinates of the image of P in the first photograph are, as before,

$$x'_1 = \frac{fx}{H - z} \tag{2.5}$$

$$y'_1 = \frac{fy}{H - z} \tag{2.6}$$

and the corresponding coordinates in the second photograph are

$$x'_2 = \frac{f(x - X)}{H - z} \tag{2.7}$$

$$y'_2 = \frac{f(y - Y)}{H - z} \tag{2.8}$$

These four equations can be solved for x, y, and z to give

$$x = \frac{x_1' X}{x_1' - x_2'} \tag{2.9}$$

$$y = \frac{y_1' Y}{y_1' - y_2'} \tag{2.10}$$

and

$$z = H - \frac{fX}{x_1' - x_2'} = H - \frac{fY}{y_1' - y_2'} \tag{2.11}$$

This completes the demonstration, by showing that the object's coordinates can be retrieved from the two sets of image coordinates, provided that the camera height H, focal length f, and the *baseline* (X, Y) are all known. The choice of baseline length is a compromise. If it is too short, the stereoscopic effect is small and heights cannot be determined accurately. On the other hand, if it is too large, the area of overlap between the two photographs will be small. The normal arrangement is to choose the baseline such that the overlap is about two thirds of the coverage of a single image. For a typical metric camera using a typical mapping film, this gives an accuracy of roughly $H/1000$ in the measured heights.

In practice, pairs of stereophotographs are analyzed in a stereoplotter, an optomechanical device which allows the operator to identify conjugate (corresponding) points in the two images and performs the calculations of equations (2.9) to (2.11) automatically. Pairs of stereophotographs can also be interpreted qualitatively by using a *stereo viewer*, which is a device that simply ensures that the operator sees one photograph with one eye and the other photograph with the other eye. The different viewpoints presented to the two eyes trick the brain into seeing an impression of relief. Recently, increasing use has been made of computer-based stereo matching in which the photographs are first scanned to produce digital images. The computer algorithm then attempts to match corresponding patches of the two images and hence to implement equations (2.9) to (2.11).* A number of commercial image processing software packages include this function. Once the heights z have been determined, the photograph can be corrected for relief displacement effects. Again, this can be performed optomechanically, using an *orthophotoscope*, or by computer processing. The result is termed an *orthophoto* or *orthophotograph*.

---

*In fact the stereo matching algorithm is more complicated than implied by these equations, since it also has to take into account the possibility of misalignment of the cameras.

2.2.5 EXAMPLES

There is a wide variety of airborne mapping cameras. A typical instrument might have a film format of $230 \times 230\,mm$, and operate at a focal length of 150 mm (though the range from about 90 to 300 mm is commonly used). The film is in the form of a roll, typically 150 m long and hence capable of recording about 600 photographs. At an altitude of 3000 m (this is roughly the greatest altitude possible without pressurizing the aircraft) this would give a coverage of $4.6 \times 4.6\,km$, a horizontal resolution of typically 0.4 m, and a vertical resolution for stereophotography of typically 3 m. Color Figure 2.1 (see color insert following page 108) shows an example of a typical aerial photograph recorded by a mapping camera.

Photography from spaceborne platforms is rather uncommon, partly because of the difficulty of retrieving exposed film from unmanned spacecraft and partly because of the advent of high-quality electro-optical systems (Section 2.3). Metric cameras have been flown on short-duration space shuttle missions, on Russian Kosmos satellites, and on the space station Mir. These have used formats of 230 to 300 mm and focal lengths of 300 to 1000 mm, giving horizontal resolutions of the order of 10 m (3 m in the case of Kosmos photography using the KVR camera systems). The largest publicly available collection of spaceborne photographs consists of nearly 900,000 U.S. military reconnaissance photographs collected by "Keyhole" cameras (Figure 2.5) on the Corona, Argon, and Lanyard satellite programs between 1959 and 1972.* The archive includes both reconnaissance and metric camera photographs. These photographs were declassified in 1995, and are not expensive.

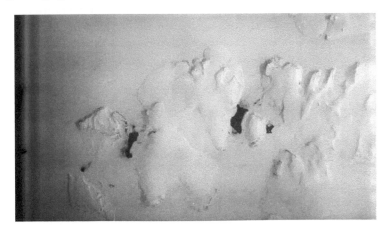

FIGURE 2.5 Keyhole-7 satellite reconnaissance photograph showing part of the Franz Josef Land archipelago, 30 April 1965. The area covered by the photograph is approximately $40 \times 22\,km$. The spatial resolution of the original is about 1 m.

---

*http://edc.usgs.gov/products/satellite/corona.html.

2.2.6 SUMMARY

Photography from the air is a comparatively simple, well understood, and widely available technique for mapping and reconnaissance. Aerial photographs are relatively straightforward to interpret in an intuitive manner. Metric cameras can produce photographs with a high degree of geometric accuracy, suitable for cartographic applications, although if significant relief is present it must be allowed for. The availability of aerial photography from space is limited, especially metric-camera photography, although there exists a very large public archive of reconnaissance photographs from the 1960s.

Since photography is a passive technique operating in visible and near-infrared regions of the electromagnetic spectrum, it can be used only during daylight. Photographs cannot be acquired through clouds.

Aerial photography still finds considerable use in remote sensing of the cryosphere, for two-dimensional and three-dimensional mapping.

2.3 ELECTRO-OPTICAL SYSTEMS IN THE VISIBLE AND NEAR-INFRARED REGION

Electro-optical systems operating in the visible and near-infrared (VIR) region possess many similarities to aerial photography. The spectral range is similar, and the instruments are again passive imaging instruments. Indeed, with the rapidly increasing popularity of digital photography, this approach can be thought of as a form of aerial digital photography. The principal difference between the electro-optical systems and their photographic counterparts is the detection mechanism. In the former case, electromagnetic radiation is detected electronically, instead of photochemically as with photography. The electronic detection can take a number of forms, including the use of photomultipliers, vidicons, semiconductor photodiodes, and one- or two-dimensional arrays of photodiodes constituting *charge-coupled devices* (CCDs), now familiar from their role in digital cameras and video cameras.

Electronic detection offers several advantages over traditional photography. The detectors can be calibrated, so that there is a known quantitative relationship between the recorded signal and the intensity of the electromagnetic radiation. The output can be digitized and used to modulate a radio signal so that the data can be transmitted to a remotely located receiver, which is a major advantage in the case where the instrument is carried on an unmanned spacecraft. The digital output data can conveniently be imported into and processed in a computer.

Although VIR imagers are used both from aircraft and from space, the spaceborne applications tend to dominate. This section will therefore concentrate primarily on the latter.

2.3.1 SCANNING GEOMETRIES

The way in which a VIR imager builds up an image of the Earth's surface depends on the configuration of its detectors. In the conceptually simplest arrangement, the incoming radiation is focused onto a two-dimensional CCD array, which functions analogously to a photographic film. The entire scene is imaged instantaneously. This mode of operation can be termed *step-stare* imaging, since the instrument "stares" at the scene, then moves on to stare at the next scene, and so on. This is still an uncommon form of imaging, largely because of the difficulty in calibrating the millions of detector elements in the array. Rather more common is *pushbroom* imaging, in which the incoming radiation is focused onto a one-dimensional CCD array. The instrument thus views instantaneously a narrow strip of the Earth's surface. Two-dimensional scanning is achieved through the motion of the platform (the aircraft or spacecraft that carries the instrument) relative to the Earth's surface. Clearly the orientation of the imaged strip must be perpendicular to this direction. This type of scanning, which has the advantage that the number of detector elements that need to be calibrated is measured in thousands rather than millions, is used in, for example, the spaceborne SPOT HRV and ASTER instruments.

The commonest form of scanning is still *whiskbroom* imaging. In this case, a single detector element (photodiode, photomultiplier, etc.) views the incident radiation. The *instantaneous field of view* (IFOV) is scanned perpendicularly to the platform motion by a rotating or oscillating mirror within the instrument, while scanning in the forward direction is again achieved through the platform's forward motion. This system, used for example by the Landsat spaceborne instruments, is illustrated schematically in Figure 2.6. The

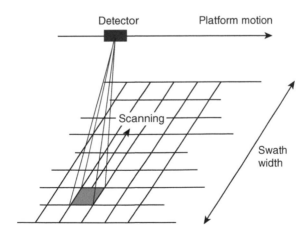

FIGURE 2.6 Whiskbroom scanning (adapted from Rees 2001). The instantaneous field of view (shaded) is scanned from side to side by a mirror.

disadvantage of including a mechanical element is the instrument is offset by the fact that there is only one or a small number of detectors to calibrate.

Finally, there is an important class of spaceborne VIR imager that does *not* move relative to the Earth's surface and so cannot make use of the relative motion of the platform. Such imagers are carried on the geostationary meteorological satellites, e.g., METEOSAT and GOES. In this case, the satellite spins about an axis parallel to the Earth's polar axis to achieve scanning in the east–west direction. North–south scanning is achieved using a slowly rotating mirror within the instrument. A typical instrument will scan the Earth's surface in a few thousand east–west scan lines and spin at about 100 rpm, giving a total time of a few tens of minutes to image the entire surface. Although geostationary imagers view a huge area of the Earth's surface, they are not particularly useful for the polar regions. This is because the line of sight from the surface to the satellite makes a small angle with the Earth's surface for latitudes less than 30° from the poles, so the image is heavily foreshortened and the radiation passes along an extensive atmospheric path (see Color Figure 2.2 following page 108).

2.3.2 Spatial Resolution

The discrete detector elements in a VIR imager are analogous to the grains in a photographic film, and determine the spatial resolution in a similar manner. It is clear that the ground-plane resolution cannot be better than the size of the detector, projected through the instrument's optics onto the ground. The corresponding feature in the image is termed a *pixel*, or picture element, and the corresponding feature on the ground is a *rezel*, or resolution element (these terms are discussed in more detail in Section 3.1). For example, the ASTER instrument uses a linear CCD consisting of 5000 detectors, each 7 μm square. Incoming radiation is projected onto this CCD through optics having a focal length of 329 mm, so the angular width seen by each detector is $7 \mu m/329 \, mm \approx 21.3$ microradians. At the nominal observing height of 705 km this corresponds to a horizontal resolution (rezel size) of 15 m.

As with photographic systems, the optical components may degrade the spatial resolution. Continuing with our example based on ASTER, and considering just the diffraction element of the resolution, we can see that the objective lens of the instrument would need to have a diameter of more than about 3 cm in order not to degrade the resolution. In fact, the diameter is 8.2 cm.

2.3.3 Spectral Resolution

Most VIR imagers are multispectral, i.e., they provide output in a number of channels corresponding to different wavelength ranges. This spectral resolution is normally achieved using filters, which can give bandwidths down to a few nanometers, although values of 20 to 50 nm are more usual. Finer resolutions can be obtained using prisms or diffraction gratings to disperse the radiation.

The comparative simplicity of providing for multispectral coverage gives the whiskbroom imager a further advantage over the pushbroom or step-stare imager. In principle, it is necessary to provide the whiskbroom imager with only a single detector for each waveband, whereas a step-stare imager requires a whole CCD array.

2.3.4 GEOMETRY OF VIR IMAGERS

The majority of VIR imagers view vertically downward, toward the nadir. The angular field of view varies considerably, depending on the application. For example, the very high resolution Quickbird and Ikonos spaceborne instruments have ground coverages of only 16 and 11 km, respectively (angular coverages 2.1° and 0.9°), whereas the coarse-resolution AVHRR instrument has a swath width of 4000 km (angular swath nearly 120°). These are extreme examples: more typical is the coverage of the Landsat satellite imagers, which have ground coverages of 185 km corresponding to an angular field of view of about 15°.

VIR imagery is subject to relief displacement effects in the same way as aerial photography. We can estimate the magnitude of the effect as follows, ignoring the Earth's curvature for simplicity. From equation (2.3), we see that the image of a point located a distance x from the image's nadir point and a height z above the ground plane will be displaced by

$$\frac{fxz}{H(H-z)}$$

relative to a point at distance x and height zero. Converting this to the equivalent horizontal displacement in the ground plane, and assuming that $z \ll H$, we see that the ground-plane relief displacement is given by

$$d \approx \frac{xz}{H} \tag{2.12}$$

For example, consider Landsat 7 ETM+ imagery for which $H = 705$ km, and x has a maximum value of 130 km. The smallest rezel size achieved by this instrument is 15 m for the panchromatic band, so the amount of relief needed to displace a pixel by the width of one rezel (i.e., $d = 15$ m) is about 80 m. Thus relief displacement effects can often be ignored altogether, unless the scene contains large differences in altitude. On the other hand, the scope for stereo mapping from nadir-viewing imagery of this type is quite limited, as only rather small baselines are possible. This problem may be overcome if the imager can be steered to view away from the nadir direction so that two images can be acquired, one at nadir and one obliquely. For a baseline b and a rezel size p, the vertical resolution is of the order of pH/b. As an example, this approach is used by the ASTER instrument for which $p = 15\,\text{m}$ and $b/H = 0.6$, giving a vertical resolution of the order of 25 m.

2.3.5 EXAMPLES

Table 2.1 summarizes the characteristics of a representative selection of the wide variety of spaceborne VIR imagers, from 1972 onward. For a more comprehensive list up to the mid- or late-1990s the reader is recommended to consult Kramer (1996) or Rees (1999).

The columns in Table 2.1 show the name of the instrument, the satellites on which it is or has been carried, the years for which data are available, the spatial resolution, and the corresponding wavebands (thermal infrared wavebands are included in this column). A figure in square brackets in this column indicates that there are several discrete wavebands within the indicated range. The next three columns show the nominal altitude and the swath width of the instrument, and whether it has a stereo capability through off-nadir pointing (this allows the imagery to be used to derive surface topography in a manner similar to stereophotography). The column "mean revisit interval" shows approximately the mean time between successive observation opportunities for a particular location rather than the repeat period of the orbit, calculated for a nominal latitude of 70°. The next column shows the maximum latitude, north and south, that can be imaged by the instrument. The figures in this column allow for both the inclination of the satellite orbit and the swath width of the sensor. The last two columns give an approximate indication of the (maximum) commercial cost of the data in 2002.

Table 2.1 illustrates a number of important points. Spatial resolutions vary from under a meter to over a kilometer, although the highest spatial resolutions come at the expense of narrow swath widths and long revisit intervals, and have become available only recently. They also tend to be expensive. Spectral resolution and number of spectral bands have generally increased over time. The table includes a few older sensors to illustrate the continuity of datasets. The last row of Table 2.1 refers to sensors carried on board geostationary satellites. Although these provide the highest temporal resolution of all spaceborne systems, they are largely unsuited to observations of the polar regions because of their imaging geometry. This is illustrated in Color Figure 2.3 (see color insert). Figure 2.7 shows an extract of typical low-resolution imagery (in fact a thermal infrared image, although the spatial resolution properties are very similar to those of the corresponding VIR imagery), Color Figure 2.4 (see color insert) shows an extract of intermediate-resolution imagery, and Figure 2.8 shows an extract of high-resolution imagery.

The availability of archived satellite images can usually be readily established from on-line catalogs. At the time of writing, some of the major catalogs are:

- NASA Earth Observing System Data Gateway at http://edcims www.cr.usgs.gov/pub/imswelcome/
- ESA's Open Distributed Information and Services for Earth Observation at http://odisseo.esrin.esa.it/
- Archives of AVHRR imagery at resolutions of 8 km at http:// daac.gsfc.nasa.gov/data/dataset/AVHRR/01_Data_Products/index. html and at 1 km at http://edcdaac.usgs.gov/1KM/1kmhomepage. html
- Sirius catalog of SPOT data at http://www.spot.com/home/sirius/ sirius.htm

However, web addresses may change and the reader is recommended to use a good web search engine to locate these and other catalog sites.

2.3.6 SUMMARY

Spaceborne electro-optical imagery in the visible and near-infrared regions combines a number of the characteristics of aerial photography with the advantages of calibrated digital data, wide spatial coverage, and rapid acquisition times, and opportunities for comparatively frequent revisiting of particular locations. The technique offers a greater diversity of spatial resolution, coverage, and spectral resolution than aerial photography. Imagery is available from the early 1970s to the present day.

Like aerial photography, this is a passive technique, operating in the same regions of the electromagnetic spectrum, and so it can be used only during daylight. Images cannot be acquired through clouds, and cloud cover is a frequent occurrence in some of the polar regions (Marshall, Rees, and Dowdeswell 1993). Another limitation arises from the fact that some spaceborne electro-optical sensors can "saturate" at the high radiances caused by reflection of sunlight from high-albedo surfaces such as snow (Dowdeswell and McIntyre 1986; Orheim and Lucchitta 1988; Hall et al. 1990; Winther 1993). The maximum radiance to which the sensor can respond is exceeded, meaning that quantitative data are lost.

2.4 THERMAL INFRARED SYSTEMS

Thermal infrared radiation (TIR) was defined in Section 2.1 as having a wavelength between about 8 μm and about 14 μm. This contains the major part of the black-body radiation emitted by objects at typical terrestrial

TABLE 2.1

Examples of Spaceborne Electro-optical VIR Imaging Instruments (See the Text for a Detailed Explanation)

Instrument	Satellites	Years	Spatial Res. (m)	Wavebands (μm)	Altitude (km)	Swath Width (km)	Stereo	Mean Revisit Interval	Max. Latitude (Deg.)	Cost per Scene (U.S. $ 2002)	Cost per Square km (U.S. $ 2002)
Quickbird	Quickbird	2001–	0.7 2.8 2.8 2.8 2.8	0.45–0.90 0.45–0.52 0.52–0.60 0.63–0.69 0.76–0.90	450	16.5	Yes	2 mo	82.9	6000	22.5
PAN	IRS 1 C/D	1996–	5	0.50–0.75	817–874	70		2 wk	81.6	2500	0.51
ETM+ (Enhanced Thematic Mapper +)	Landsat 7	1999–	30 30 30 30 30 60 30 15	0.45–0.52 0.52–0.60 0.63–0.69 0.76–0.90 1.55–1.75 10.42–12.50 2.08–2.35 0.52–0.90	705	185		1 wk	82.6	600	0.02
ASTER	Terra	2000–	15 15 15 15 30 30 90	0.52–0.60 0.63–0.69 0.76–0.86 0.76–0.86 1.60–1.70 2.15–2.43 [5] 8.13–11.65 [5]	705	60	Yes	2 wk	82.1	100	0.03
Thematic Mapper (TM)	Landsat 4, 5	1982–	30 30 30 30 30 120 30	0.45–0.52 0.52–0.60 0.63–0.69 0.76–0.90 1.55–1.75 10.42–12.5 2.08–2.35	705	185		2 wk	82.6	400–1500	0.01–0.05

Multispectral scanner (MSS)	Landsat 1–5	1972–1993	80 80 80 80	0.50–0.60 0.60–0.70 0.70–0.80 0.80–1.10	920, 705	185	2 wk	82.6	200	0.006
WiFS	IRS 1 C/D	1996–	188 188	0.62–0.68 0.77–0.86	817–874	812	1 d	85.0	800	0.001
MODIS	Terra, Aqua	2000–	250 250 500 500 500 500 500 1000	0.62–0.67 0.84–0.88 0.46–0.48 0.55–0.57 1.23–1.25 1.63–1.65 2.11–2.16 0.41–14.39 [28]	705	2330	12 h	90	0	0
AVHRR 3	NOAA 15, 16	1998–	500 1000 1000 1000 1000 1000	0.58–.68 0.72–1.00 1.58–1.64 3.55–3.93 10.3–11.3 11.5–12.5	840	3000	8 h	90	100	0.00002
AVHRR 1	NOAA 6, 8, 10, 12	1979–1994	1000 1000 1000 1000 1000	0.58–0.68 0.73–1.10 3.44–3.93 10.5–11.3 10.5–11.3	840	3000	8 h	90	100	0.00002
VISSR	Meteosat, GOES, GMS	1977–	2500 5000 5000	0.5–0.9 5.7–7.1 10.5–12.5	35,800	Earth disk	0.5h	60	0	0

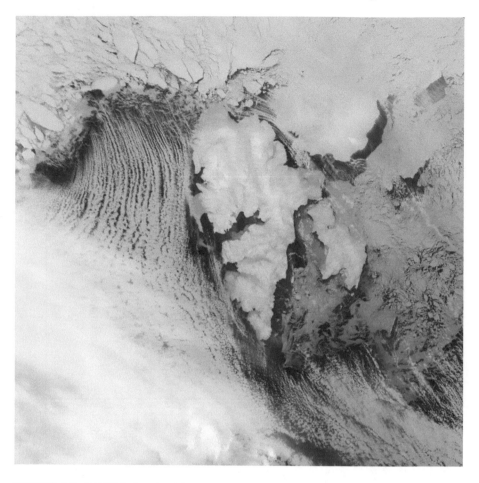

FIGURE 2.7 AVHRR band 4 (thermal infrared) image showing Svalbard, 2 December 1994. (Image reproduced by courtesy of NERC Satellite Receiving Station, University of Dundee.)

temperatures, and the principal reason for wanting to detect this radiation is to measure Earth surface temperatures or to deduce surface emissivities.*

Although the general principles of detection and scanning are similar to those for the VIR devices discussed in Section 2.3, the longer wavelength and hence lower photon energies of TIR radiation do introduce some complications. Photodiodes suitable for detecting TIR radiation have to be constructed from more exotic semiconductor materials, and while it is possible to construct two-dimensional arrays of detectors, these do not tend to have as many

---

*That is, in the context of this book. Detection of TIR also plays a major part in atmospheric remote sensing, including temperature profiling, the measurement of water vapor and ozone concentrations, and determination of cloud properties. These applications are not discussed in this book.

FIGURE 2.8 Extract of Landsat 7 ETM+ band 8 imagery showing part of the Brøgger Peninsula, Svalbard. The highly crevassed terminus of the glacier Kongsvegen is visible at top right. The image covers an area of 24 × 16 km.

elements as the CCDs used in VIR imagers. Whiskbroom imaging is still the norm, and as discussed in Section 2.3.3, this has the advantage of simplifying the provision of multispectral coverage. The detector is usually cooled to reduce self-generated noise. Thermal infrared imaging is often combined with VIR imaging in the same instrument, as Table 2.1 demonstrates.

2.4.1 SPATIAL RESOLUTION

The factors that determine the spatial resolution of a TIR imager are similar to those for a VIR imager. The longer wavelength does mean, however, that the spatial resolution is usually coarser. For example, consider the ETM+ instrument carried on the Landsat 7 satellite. This operates from a nominal height of 705 km, and the optics have an equivalent focal length of 2348 mm. The detectors for the VIR bands (1–5) have a physical size of 0.1 mm, giving a spatial resolution at the ground of 30 m, while the TIR detector has a size of 0.2 mm, giving a resolution of 60 m. The diameter of the objective mirror of the optical system is 0.406 m, so the contribution from diffraction* is of the

---

*The theory of diffraction shows that the angular resolution of a simple imaging instrument with an objective diameter D, operating at a wavelength λ, cannot be better than about λ/D radians. It applies to optical, infrared, and passive microwave systems, although some of the radar systems discussed in this book circumvent the limit.

order of 30 microradians at a wavelength of 12.5 µm, equivalent to about 20 m
spatial resolution at the ground. This system is thus not far away from being
diffraction-limited, and significantly better spatial resolutions would require
larger-diameter optics as well as smaller detectors.

2.4.2 SPECTRAL RESOLUTION

The spectral resolution required for TIR imagers depends on the application.
For the applications of most relevance to this book, namely imaging of the
Earth's surface, high spectral resolution is not usually required, and bandwidths in the range 0.1 to 1 µm are common. (Profiling of atmospheric properties requires significantly higher spectral resolution.) The wavelengths of
these bands are usually around 3 or 4 µm and around 8 to 14 µm, thus avoiding
the strong water vapor absorption feature around 6 to 7 µm (Figure 2.1).

2.4.3 RADIOMETRIC RESOLUTION

Since the primary purpose of TIR images is usually to determine the brightness temperature of the incident radiation, the ability to discriminate small
changes in this quantity is of fundamental importance. This is usually specified
by the *noise equivalent temperature difference* (NEΔT), and is determined by
a variety of factors including the type of detector, its physical temperature,
and the *integration time*, i.e., the time for which the detector views a single
rezel. Typical values of NEΔT range from 0.1 to 1 K, implying that relative
differences in brightness temperature can be measured with this accuracy.

2.4.4 ATMOSPHERIC CORRECTION

Although Section 2.4.3 pointed to the desirability of measuring the brightness
temperature of the incident radiation, what we usually require is the brightness temperature of the radiation leaving the Earth's surface. As a result of
atmospheric propagation effects, these two temperatures may be of the order
of 10 K different in the case of a spaceborne observation.

There are three main techniques for atmospheric correction of TIR data
(Cracknell and Hayes 1991). The first of these is physical modeling, using one
of the atmospheric transmission models, such as LOWTRAN or MODTRAN.
Since the largest part of the correction is normally due to the water vapor
component, which is variable both spatially and temporally, this approach is
unreliable unless accurate local data on water vapor are available. Such data
can in some cases be generated from a remote sensing instrument carried on
the same platform as the TIR imager.

The second approach to atmospheric correction is normally called the
split-window method. Here, the TIR instrument makes measurements in two

closely spaced wavebands, typically around 10 and 11 μm. It is then assumed that the at-surface brightness temperature is a linear function of the at-sensor brightness temperatures measured in these two wavebands. The coefficients in this linear function have to be determined empirically, and they vary according to the type of surface and whether it is daytime or night-time, but over homogeneous areas, such as the ocean, the technique is accurate to about 0.5 K. It is very widely used, and an example is given in Section 8.5.

Finally, in the *two-look* method the instrument views the Earth's surface in two directions, typically at nadir and at a large angle off-nadir. By comparing the brightness temperatures observed in these two directions, the contribution from the atmosphere can be calculated and subtracted (Rees 2001). This method is used by the AATSR (Advanced Along-Track Scanning Radiometer) carried on the Envisat satellite.

2.4.5 EXAMPLES

Table 2.1 includes several examples of spaceborne instruments with significant coverage of the TIR region; we note particularly ASTER, MODIS, AVHRR, and VISSR. To this list can be added the class of instruments designed primarily for TIR imaging. A good example of this type of instrument is the AATSR. This has wavebands at 3.7, 11.0, and 12.0 μm, plus VIR bands at 0.65, 0.85, 1.27, and 1.6 μm. It uses a conical scanning technique to scan the IFOV between nadir and 52° forward of nadir. The rezel size at nadir is 500 m (bands up to 1.6 μm) and 1000 m for the thermal infrared bands. The AATSR instrument, which has operated since 2002, provides data continuity with the earlier ATSR, which has operated since 1991 (Figure 2.9).

2.4.6 SUMMARY

Thermal infrared imaging systems are mainly used for measuring the brightness temperature of the Earth's surface (and of cloud tops). Although there are dedicated spaceborne TIR instruments, the function is often combined with VIR imaging. Spatial resolutions are a little coarser for TIR than for VIR imaging.

Thermal infrared imaging systems do not detect reflected sunlight, so they can be used at night. However, the radiation does not penetrate through clouds.

2.5 PASSIVE MICROWAVE SYSTEMS

The thermal infrared systems discussed in Section 2.4 detect blackbody radiation in the thermal infrared (typically 3 to 15 μm) waveband. Passive microwave systems measure black-body radiation in the microwave

FIGURE 2.9 ATSR-2 image (12 μm) showing large icebergs calving from the Ross ice shelf in March 2000. Lighter shades of gray denote lower brightness temperatures. (Image reproduced with permission of Rutherford Appleton Laboratory and University of Leicester Earth Observation Science Group.)

(wavelengths typically 3 mm to 6 cm, or equivalently frequencies between 5 and 100 GHz) region. As the name implies, it is a passive technique. Since it is a microwave technique, however, it has the ability to penetrate through most clouds. It can thus be characterized as an all-weather, day-and-night, technique. As with thermal infrared systems, the purpose is to measure the brightness temperature of the incident radiation, to deduce either the physical temperature of the Earth's surface or its emissivity.

The much longer wavelengths of microwave radiation mean that the photons are very much less energetic than those of visible light (typically a few tens of micro electron volts, compared with a few electron volts), so completely different detection techniques are used. A passive microwave radiometer is effectively a radio telescope viewing downward (Figure 2.10). The incident radiation is collected by an *antenna*, which converts it into a fluctuating voltage difference that can be amplified and detected.

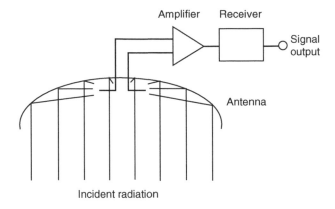

FIGURE 2.10 Passive microwave radiometer (schematic). (Adapted from Rees 2001.)

2.5.1 Spatial Resolution and Swath Width

The spatial resolution of a passive microwave radiometer is set by the diffraction limit. The angular FWHH is of the order of λ/D, where λ is the wavelength and D the width (e.g., the diameter in the case of a dish) of the antenna. The long wavelengths imply coarse angular resolutions: for an antenna 1 m in diameter operating at a wavelength of 2 cm, the angular resolution is of the order of 1°, which would give a horizontal spatial resolution of about 14 km from a typical spacecraft altitude of 700 km. This is perhaps the main disadvantage of passive microwave methods.

The large size of an antenna means that it is not practicable to construct arrays of large numbers of them in order to provide a wide spatial coverage. Instead, the *beam* of the antenna (its direction of maximum sensitivity) is scanned, by either mechanical or electrical means. The usual form of mechanical scanning is the *conical scan*, in which the beam is rotated in a wide cone about the nadir direction, as shown in Figure 2.11.

A potential disadvantage of this kind of mechanical scanning is that it may cause undesirable vibration of the instrument. An alternative form of scanning has no moving parts. The antenna consists of a closely spaced array of smaller antennas. The phases of the signals detected by these antennas can be changed under electronic control, which allows the direction of the beam to be altered. This mode of operation, often referred to as a *phased array*, is described in more detail by Rees (2001).

2.5.2 Spectral Resolution and Frequency Coverage

With the exception of instruments designed for atmospheric profiling, passive microwave radiometers do not need to achieve particularly high spectral resolution. Bandwidths are typically around 1% of the operating frequency.

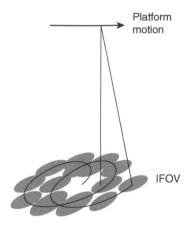

FIGURE 2.11 Conical scanning by a passive microwave radiometer (schematic).

Most passive microwave radiometers provide coverage of a number of different frequencies, often in two polarizations. For spaceborne observations, frequencies below about 5 GHz are unsuitable because of the very poor spatial resolution. The graph of atmospheric transparency shown in Figure 2.1 implies that at frequencies below about 15 GHz the detected brightness temperature will be dominated by surface emission. Atmospheric water vapor causes a correction of a few kelvin. Between about 15 and 35 GHz the surface signal still dominates, although the contribution due to water vapor is significantly larger, and observations in this frequency range are often used to provide a simple correction for this. Above about 35 GHz the effects of molecular absorption in the atmosphere become dominant, and these frequencies are more useful for atmospheric profiling than for surface imaging.

2.5.3 RADIOMETRIC RESOLUTION

As for the thermal infrared systems discussed in Section 2.4, the sensitivity of a passive microwave radiometer can most usefully be defined in terms of the detectable change in brightness temperature. This is determined by the "system noise temperature" (a function of the instrument design and its physical temperature), the integration time, and the bandwidth. Values of a few tenths to 1 K are typical.

2.5.4 EXAMPLES

Table 2.2 lists examples of spaceborne passive microwave radiometers that are not exclusively used for atmospheric sounding, i.e., those that have at least some capability for surface imaging.

TABLE 2.2
Some Passive Microwave Radiometers with Surface Imaging Capabilities

Instrument	Satellite	Years	Spatial Res. (km)	Frequency (GHz) and Polarization	Swath Width (km)	Max. Latitude (Degrees)
ESMR	Nimbus 5	1972–1976	25	19.35 H	3000	90
SMMR	Nimbus 7	1978–1988	136 × 89	6.6 H, V	780	84.2
			87 × 57	10.7 H, V		
			54 × 35	18.0 H, V		
			47 × 30	21.0 H, V		
			28 × 18	37.0 H, V		
SSM/I	DMSP	1987–	70 × 45	19.35 H, V	1400	87.5
			60 × 40	22.24 V		
			38 × 30	37.0 H, V		
			16 × 14	85.5 H, V		
AMSR/E	Aqua	2002–	74 × 43	6.93 H, V	1445	88.3
			51 × 30	10.65 H, V		
			27 × 16	18.7 H, V		
			31 × 18	23.8 H, V		
			14 × 8	36.5 H, V		
			6 × 4	89.0 H, V		

As the table indicates, the currently operational instruments are the SSM/I (Special Sensor Microwave Imager) carried on board the DMSP (Defense Meteorological Satellite Program) series of satellites (Figure 2.12), and the AMSR/E (Advanced Microwave Scanning Radiometer) carried on the Aqua satellite. Color Figure 2.3 (see color insert) illustrates the spatial resolution of the SSM/I instrument. By virtue of their wide swaths and near-polar orbits, these give coverage to within a couple of degrees of the Earth's geographical poles. The wide swaths also mean that the revisit period for a given location is short (less than one day at latitude 70°) and that global coverage can be obtained from just a few days' data. The disadvantage of the comparatively coarse resolution is thus compensated by the ability to produce frequently updated synoptic views.

2.5.5 SUMMARY

Like thermal infrared systems, passive microwave radiometers are used for measuring the brightness temperature of the Earth's surface. Compared with TIR imagers, passive microwave systems provide a substantially poorer spatial resolution but generally a very broad swath width giving synoptic, near-global coverage from a few days' data. Passive microwave radiometers usually

FIGURE 2.12 H-polarized 85 GHz SSM/I brightness temperature image of the northern hemisphere, 1 December 2003. Lighter shades of gray correspond to increasing brightness temperature. Black indicates areas of no data. (Original data obtained through EOSDIS NSIDC Distributed Active Archive Center (NSIDC DAAC), University of Colorado at Boulder.)

provide considerable diversity of frequency and polarization, and the ability to collect 7 to 10 independent variables from each rezel is not uncommon. Like TIR systems they do not detect reflected sunlight and can therefore be operated at night. Unlike TIR systems they can also be used to observe through most cloud covers.

2.6   LASER PROFILING

Laser profiling is the first and simplest of the active remote sensing techniques that we shall consider. It is primarily a ranging method designed to measure the Earth's surface topography (for example, the surface profile of the ocean or an ice sheet), and therefore not an imaging technique. Laser profilers are usually operated from low-flying aircraft, although satellite instruments do exist. The Geoscience Laser Altimeter System* (GLAS) is carried on ICESat (launched January 2003). It has a vertical resolution of about 5 cm.

The basic idea of a laser profiler is extremely simple. The instrument emits a short pulse of "light" (in fact, usually near-infrared radiation) from a downward-pointing laser. At the same instant, an electronic clock is started. The pulse propagates down through the atmosphere, bounces off the Earth's surface, propagates back up through the atmosphere, and is detected by a photodiode. Detection of the pulse stops the clock, so that the two-way travel time to the surface can be deduced. If the propagation speed is known, the range to the surface can be determined. If the absolute position of the instrument is known, the absolute position of the reflecting point on the Earth's surface can therefore also be determined.

2.6.1 SPATIAL RESOLUTION

The spatial resolution of a laser profiler has two aspects: horizontal resolution and vertical (height) resolution. The horizontal resolution is determined fundamentally by the beam width of the laser. For a laser with a beam width of $\Delta\theta$ (radians) operating from a height H above the surface, the horizontal resolution is clearly

$$\Delta x = H\Delta\theta \qquad (2.13)$$

For example, a system for which $\Delta\theta = 0.5$ milliradian operated from $H = 200$ m would illuminate a region on the Earth's surface of width $\Delta x = 0.1$ m. This is often referred to as the laser's *footprint*.

The vertical resolution is determined by the accuracy with which the two-way travel time can be measured. Assuming that the electronic clock is sufficiently accurate, the accuracy with which the travel time can be measured is governed by the *rise time* t_r of the detected pulse (the time it takes to increase from zero to maximum power) and its *signal-to-noise ratio* S. For a single pulse, the vertical resolution is given by

$$\Delta z = \frac{ct_r}{2S} \qquad (2.14)$$

where it is assumed that the pulses travel at the speed of light c. However, laser profilers are pulsed systems and if the *pulse repetition frequency* (PRF) f is high

---

*http://glas.gsfc.nasa.gov/.

enough and the platform speed v is low enough, a given footprint-sized area on the Earth's surface will be sampled several times. Since each of these samples is independent, and there are $(fH\Delta\theta/v)$ of them, the vertical resolution when these samples are averaged becomes

$$\Delta z = \frac{ct_r}{2S}\sqrt{\frac{v}{fH\Delta\theta}} \qquad (2.15)$$

For example, suppose the system that we have just considered is operated from a platform with a speed of $50\,\mathrm{m\,s^{-1}}$ and gives a rise time t_r of $5\,\mathrm{ns}$ with a signal-to-noise ratio of 1 and a PRF of $2000\,\mathrm{s^{-1}}$. While (2.14) shows that a single pulse will give a vertical resolution of $0.75\,\mathrm{m}$, (2.15) shows that averaging of four pulses over a single footprint can reduce this to $0.38\,\mathrm{m}$.

Equation (2.15) indicates the features of a laser profiler's design that will give it a high vertical resolution. The rise time of the returned pulse should be as short as possible. This obviously indicates that the transmitted pulse should be as short as possible, but reflection from a rough surface will spread the reflected pulse over time. The signal-to-noise ratio should be as high as possible, which indicates that a large transmitted power is needed, and also that the height H should be small to reduce losses through absorption and geometrical spreading. The platform speed should be small and the horizontal resolution $H\Delta\theta$ large, though these considerations are clearly not optimal for other reasons. Finally, the PRF should be as large as possible.

In fact, there is an upper limit on the PRF. If it is too large, the measurement of the range H becomes ambiguous because it is no longer possible to identify which return pulse corresponds to which transmitted pulse. The simplest situation is that in which the second pulse is not transmitted until the first pulse has been received, and this implies that

$$f < \frac{c}{2H} \qquad (2.16)$$

For airborne systems this implies that the PRF should be less than a few tens or hundreds of thousands of pulses per second, and for spaceborne systems it should be less than a few hundred pulses per second. Actual systems are generally operated well within these limits.

Airborne laser profiling systems can be scanned. This is achieved using an oscillating mirror to steer the beam from side to side, while the forward motion of the aircraft provides scanning in the forward direction. In a typical configuration, a scanning airborne laser profiler has a PRF of $30\,\mathrm{kHz}$ and is flown at a height of $1000\,\mathrm{m}$ and a speed of $70\,\mathrm{m\,s^{-1}}$. If the mirror scans at $35\,\mathrm{Hz}$ through an angle of $20°$ either side of nadir, this gives a swath width of about $480\,\mathrm{m}$ with a mean spacing of the sampled points of $1\,\mathrm{m}$ in both the forward and side-to-side directions.

2.6.2 ATMOSPHERIC CORRECTION

In the preceding sections it was assumed that the pulses travel at the speed of light c. In fact, they will travel somewhat more slowly than this. For electromagnetic radiation making a one-way journey of length L through the atmosphere, the propagation time T can be written as

$$T = \frac{L + P}{c} \tag{2.17}$$

where P is the propagation delay expressed as a distance rather than as a time. P is thus the amount that should be subtracted from the apparent one-way range calculated on the assumption that the pulse propagated at the speed of light c.

The propagation delay P has two major components, arising from the dry atmosphere and from water vapor, respectively. The dry atmosphere component is proportional to the number of *air masses* that the radiation has passed through. For a vertically propagating pulse, this is given by the difference between the atmospheric pressure at the bottom and top of the path, divided by the standard sea-level atmospheric pressure of 101.325 kPa. Thus a pulse from a satellite, where the atmospheric pressure is essentially zero, propagating all the way down to sea level will pass through an air mass of 1, whereas a pulse an aircraft at 1000 m will pass through about 0.1 air masses. The coefficient of proportionality is about 2.4 m per air mass at a wavelength of 1 μm.

The propagation delay due to water vapor depends on the column-integral of the water vapor density along the propagation path. This quantity is normally expressed as the equivalent depth of precipitable water, i.e., the depth of the layer of water that would be produced by condensing the vapor. Densities of water vapor are highly variable both spatially and temporally, and typical values of this quantity for the entire atmosphere range between about 5 and 200 mm. The coefficient for the propagation delay is about 0.4 m per meter of precipitable water at a wavelength of 1 μm.

2.6.3 EXAMPLES

We have already illustrated typical values for a generic airborne laser profiler. Data produced from a typical scanning laser profiler are illustrated in Figure 2.13. Laser profilers have only recently emerged as a spaceborne technology. The first was the Balkan-1 instrument carried on the Mir space station from 1995. A second spaceborne laser profiler, the GLAS (Geoscience Laser Altimeter System, also known as the GLRS) is carried on the ICESat satellite.* The parameters for this instrument are $H \approx 600\,\text{km}$, $\Delta\theta \approx 0.12$

---

*http://icesat.gsfc.nasa.gov/intro.html.

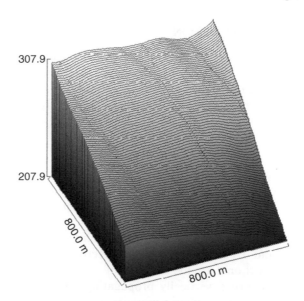

FIGURE 2.13 Visualization of the surface topography of an 800×800 m region of the glacier Midre Lovénbreen, Svalbard, obtained from a scanning laser profiler. Note that the lines do not represent actual scan lines of the profiler. They are included as a way of visualizing the topography, on which several meltwater channels are clearly visible. (Data from author's research in collaboration with Dr N. S. Arnold, supplied by NERC Airborne Remote Sensing Facility.)

milliradian, $f = 40\,\text{s}^{-1}$. Equation (2.13) shows that the horizontal resolution is about 70 m. Since the satellite travels approximately 200 m between pulses, each footprint is sampled only once, so the vertical resolution is given by equation (2.14). The values of t_r and S to be achieved over a smooth ice surface are such that the value of Δz should be around 0.1 m. ICESat was launched on 12 January 2003 and began collecting data from the GLAS system in February 2003. However, the laser stopped working on 29 March after the instrument had collected 36 days of data. ICESat carries three lasers, and at the time of writing, further datasets have been collected using lasers 2 and 3.

During the main mission of ICESat it will be placed in an exactly repeating orbit, making 2723 orbits in a 183-day cycle. This means that complete near-global coverage (to latitude 86°) can be obtained twice per year. This coverage consists of about 600 million range values with an along-track spacing of 200 m and an across-track spacing of 0.13° of longitude (about 5 km at latitude 70°).

2.6.4   SUMMARY

Laser profiling is an active technique for measuring the height profile of the Earth's surface or (if they are present) the tops of clouds. It can achieve vertical resolutions of the order of 0.1 m. It is at present almost entirely an airborne technique, although the GLAS instrument has been carried on board the ICESat satellite since early 2003. Data rates from laser profilers are low, at a few tens of thousand samples per second or less.

2.7 RADAR ALTIMETRY

The radar altimeter is conceptually very similar to the laser profiler. Like the laser profiler, its purpose is to measure the range to the Earth's surface by measuring the two-way travel time of a short pulse of radiation. As its name implies, the radar altimeter uses microwave radiation, with frequencies typically around 10 GHz. This means that, unlike laser profiling, it can observe through cloud. Most other differences between the characteristics of the two types of instrument arise from the much longer wavelength at which the radar altimeter operates.

Figure 2.14 shows very schematically the operation of a radar altimeter. A trigger signal starts the clock and causes a very short pulse to be sent to the antenna. This pulse travels down to the Earth's surface, is reflected, and is collected by the same antenna. A switch directs the pulse to the receiver. This resolves the temporal structure of the returned pulse by measuring the power received in successive short time intervals. The output is thus characterized by both the time delay for the two-way propagation to the Earth's surface and back, and the time structure of the returned pulse, usually referred to as the *waveform*.

2.7.1 SPATIAL RESOLUTION

As for the laser profiler, we are concerned with both the horizontal and the vertical spatial resolution. Paradoxically, the beam width of the antenna does not normally play an important part in determining these resolutions. To see why this is so, it is helpful to develop a very simple model of radar altimetry

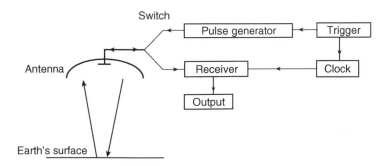

FIGURE 2.14 Operation of a radar altimeter (schematic).

(Rees 2001). We assume that the Earth's surface is flat and that it consists of a uniform density of incoherent scatterers. We also ignore the effects of geometric spreading and the antenna's finite beam width (i.e., we assume that this is very large). These assumptions mean that if the antenna were to transmit continuously, the received power would just be proportional to the illuminated area of the Earth's surface.

In order to consider the effect of transmitting a pulse, we suppose that it is uniform and of duration t_p, beginning at time zero. This can be visualized as a "scattering zone" of thickness $ct_p/2$ propagating at speed $c/2$ away from the antenna, so that any scatterer within this zone at time t contributes to the power received at time t. The factors of 1/2 allow for the two-way journey of the pulse.

Evidently, no return signal is received until $t = 2H/c$, where H is the height of the antenna above the Earth's surface. A time Δt later than this, where Δt is less than t_p, the scattering zone intersects the surface in a disk of radius r, as shown at the left of Figure 2.15. This disk expands over time until $t = 2H/c + t_p$, when the trailing edge of the scattering zone just touches the surface. At $t = 2H/c + \Delta t$, where Δt is greater than t_p, the scattering zone intersects the surface in an annulus with inner radius r_1 and outer radius r_2, as shown at the right of Figure 2.15. It is not difficult to show that in the first case

$$r \approx \sqrt{cH\Delta t} \qquad (2.18)$$

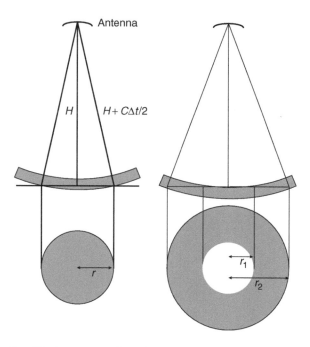

FIGURE 2.15 Simplified geometry of a radar altimeter pulse. (Adapted from Rees 2001.)

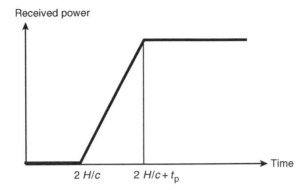

FIGURE 2.16 Waveform from a flat surface, using the simple model developed in the text.

while in the second case

$$r_1 \approx \sqrt{cH(\Delta t - t_p)} \qquad (2.19)$$

and

$$r_2 \approx \sqrt{cH\Delta t} \qquad (2.20)$$

Equation (2.18) shows that the illuminated area, and hence the received power, increases uniformly with time until $t = 2H/c + t_p$. Equations (2.19) and (2.20) show that at times later than this the illuminated area, and hence the received power, is constant (Figure 2.16).

It is clear that useful information about the surface is obtained only while the received power is changing. The effective horizontal resolution of the altimeter is thus determined by the maximum size of the illuminated disk, just before it becomes an annulus. The horizontal resolution is thus, from equation (2.18),*

$$\Delta x = 2\sqrt{cHt_p} \qquad (2.21)$$

We can now consider whether the assumption that the antenna's beam width is very large is a reasonable one. If the FWHH of the beam (given approximately by λ/D where D is the antenna diameter) is $\Delta\theta$, the arguments that we have just developed are valid provided that

$$H\Delta\theta \gg 2\sqrt{cHt_p} \qquad (2.22)$$

---

*This can be corrected for the fact that the Earth is not flat by replacing H with an effective height H', where $1/H' = 1/H + 1/R$ and R is the Earth's radius of curvature.

This condition, which is virtually always realized in practice, is referred to as *pulse-limited operation* of the altimeter.

The vertical resolution of the altimeter is found using equation (2.14). The rise time of the returned pulse is just t_p, and the signal-to-noise ratio S is about 1. (This is because the assumption that the scatterers are incoherent is not correct, and the waveform shown schematically in Figure 2.9 should actually have noise due to interference superimposed upon it.) Thus, very simply,

$$\Delta z = \frac{ct_p}{2} \tag{2.23}$$

for a single pulse. In practice, many pulses are averaged to improve the vertical resolution.

2.7.2   INFORMATION IN THE WAVEFORM

The theory in the preceding section was developed for a flat surface. If the surface is rough on a scale shorter than the horizontal resolution Δx, the return signal will be broadened in time by an amount corresponding to the roughness. It will no longer have the simple ramp structure shown in Figure 2.16, and the effective rise time will be increased to t_p', where

$$t_p'^2 = t_p^2 + \frac{\sigma_h^2}{c^2} \tag{2.24}$$

in which σ_h^2 is the variance of the surface height.* Thus the waveform contains useful information about the surface roughness, and is extensively used for estimating the *significant wave height* over oceans. It can also indicate if the scattering does not all take place from the surface but contains a significant proportion of volume scattering. This can occur over terrestrial ice masses; if the surface snow is very dry, the microwave radiation from the altimeter can penetrate it to a significant extent (see Section 4.2.6). A third case in which the waveform can provide useful information is if the scattering contains a strong specular component, for example, from a large, smooth floe of sea ice. In this case the waveform is dominated by a pronounced spike.

2.7.3   RESPONSE TO SLOPES

If we consider a radar altimeter looking vertically downward at a sloping plane (Figure 2.17), it is evident that the altimeter will measure the distance to the

---

*This also implies that the horizontal resolution is coarsened over a rough surface, from equation (2.21).

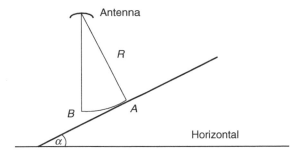

FIGURE 2.17 Illustration of slope-induced error.

point A that is closest to the antenna. (This will be true only if the slope angle α is less than about half the antenna's beam width, so that a significant amount of power can be returned to the antenna from points in the vicinity of A.) However, there is no way of telling from a single observation that this is what has happened, and the position of the reflecting point will be incorrectly deduced to be B, at the same distance R from the antenna but vertically below it. The distance between A and B is termed the *slope-induced error*, and simple geometry shows that it is approximately $R\alpha$ horizontally and $R\alpha^2/2$ vertically, provided that the angle α (which is measured in radians) is small. Since the mean slope of an ocean surface is usually less than 10^{-4} radians, this effect can be ignored even in a spaceborne observation. On the other hand, a slope of only 0.01 radian (about half a degree) gives errors of typically 8 km horizontally and 40 m vertically for a spaceborne measurement. These slope-induced errors can be corrected in many cases, by modeling the surface topography and using spatially adjacent altimeter measurements to fit the parameters of the model.

2.7.4 RESPONSE TO ABRUPT CHANGES IN HEIGHT

The previous section illustrated a problem that arises in the case of smooth but sloping terrain. A more serious problem can arise if the terrain contains abrupt changes in height, most commonly in the transition from sea to land or over rugged terrain. This is a technological limitation arising from the way the waveform is sampled. For example, suppose the altimeter is carried on a satellite at a height of 800 km above the surface, and that the waveform is sampled at intervals corresponding to a vertical resolution of 0.1 m. If the receiver is switched on at the instant that the pulse is transmitted, it will record about 16 million samples before the first return signal is received. Since pulse repetition frequencies of $100\,\mathrm{s}^{-1}$ are typical of spaceborne instruments, this would require a mean data rate of at least 1.6×10^9 samples per second. To reduce this requirement very substantially, the receiver is not switched on immediately the pulse is transmitted. The instrument is programmed to predict ("track") when the next pulse will arrive, based on the delays of previous pulses, and only switched on just before the expected arrival time. It is evident that abrupt changes in surface height may outwit this procedure, in a phenomenon known as "loss of lock" or "loss of tracking." One way of reducing this problem is to vary the temporal resolution of the instrument according to the type of surface that it is viewing, with the finest resolutions being used for the smoothest surfaces.

2.7.5 Atmospheric Correction

Very similar remarks can be made for atmospheric correction of radar altimeter measurements as for laser profiler measurements, and equation (2.17) again provides a suitable way in which to specify the magnitude of the effects. The value of P due to the dry atmosphere is 2.33 m per air mass, and that due to water vapor is 7.1 m per meter of precipitable water. Both of these are independent of the frequency of the radiation. However, for spaceborne observations there is a third important term, due to the ionosphere. This depends on the frequency f of the transmitted radiation, and also on the *total electron content* N_t of the ionosphere, defined as the column-integral of the number density of electrons. In numerical terms, the expression for P is

$$\left(\frac{P}{m}\right) = 4.0\left(\frac{N_t}{10^{17}\,m^{-2}}\right)\left(\frac{f}{GHz}\right)^{-2} \tag{2.25}$$

Thus at a typical frequency of 13.6 GHz and for a typical daytime total electron content of $3 \times 10^{17}\,m^{-2}$, P would be about 0.06 m. Some radar altimeters make observations at two frequencies so that the ionospheric effect can be identified and subtracted.

2.7.6 Examples

The concept of spaceborne altimetry was first demonstrated by the Skylab mission in 1973. Table 2.3 summarizes the principal spaceborne radar altimeters. The spatial resolutions include the effects of onboard averaging of pulses. As the table indicates, continuous near-global coverage, with a vertical resolution of a few centimeters, has been available since the launch of ERS-1 in 1991.

2.7.7 Summary

Radar altimetry is, like laser profiling, an active technique for measuring the Earth's surface topography (Figure 2.18). Unlike laser profiling, it is not

TABLE 2.3
Principal Spaceborne Radar Altimeters

Satellite	Years	Frequency (GHz)	t_p (ns)	Spatial Resolution (m) Horizontal	Spatial Resolution (m) Vertical	Height (km)	Max. Latitude (degrees)
GEOS-3	1975–1978	13.9	15	7000	0.2	843	65
Seasat	1978	13.5	3.1	7000	0.1	800	72
Geosat	1985–1990	13.5	3.1	6700	0.05	800	72
ERS-1, 2	1991–	13.8	3.0	3500	0.05	800	81.5
			12.1				
Topex-Poseidon	1992–	13.6	3.1	2200	0.02	1336	66
		5.3					
Mir	1996–2001	13.8	3.0	2300	0.1	390	52
Jason-1	2001–	13.6	3.1	2200	0.02	1336	66
		5.3					
Envisat	2002–	13.6	3.1	1700	0.05	800	81.5
		3.2	6.3	2400	0.2		

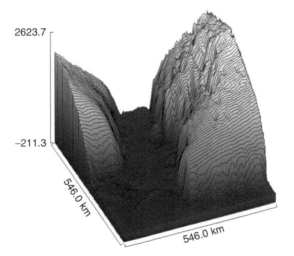

FIGURE 2.18 Visualization of the topography of the Antarctic ice sheet from GEOSAT radar altimeter data (Herzfeld and Matassa 1999) gridded at a spacing of 3 km. The image shows parts of Mackenzie Bay and the Amery ice shelf.

affected by cloud cover. Spaceborne radar altimetry can achieve vertical resolutions of a few centimeters with a corresponding horizontal resolution of a few kilometers. It is also capable of measuring surface roughness. Corrections are required for observations over sloping terrain, and performance can be significantly degraded where the surface slope exceeds a few degrees or where there are abrupt changes in surface height.

2.8   RADIO ECHO-SOUNDING

The last of the ranging systems that we will discuss is radio echo-sounding. This is a technique for measuring the thickness of glaciers and ice sheets. It relies on the fact that freshwater ice is notably transparent to radio frequency radiation in the VHF band (i.e., around 100 MHz), as discussed in Section 4.3. Like the two preceding methods, the technique involves transmitting a short pulse of radiation and measuring the two-way travel time for the reflection from (in this case) the underlying bedrock (Figure 2.19). Reflections are also

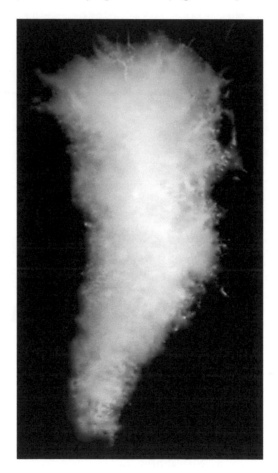

FIGURE 2.19 Thickness of the Greenland ice sheet derived from radio echo-sounding measurements (Bamber, Layberry, and Gogineni 2001a, b). The data have been gridded at intervals of 5 km and cover an area of 1505 × 2805 km.

obtained from the ice surface and from internal layers if they show sufficiently strong dielectric contrast. Range (thickness) resolutions of the order of 1 m are possible.

Because of the long wavelengths of VHF radiation (e.g., 3 m in free space at 100 MHz) it is not feasible to construct antennas with narrow beam widths. In order not to compromise the spatial resolution, radio echo-sounding is thus primarily an *in situ* technique, the equipment being located either on the ice surface itself or in a low-flying aircraft.

Other similar techniques also exist. *Ground-penetrating radar*, when used for measurements of glaciers, snow fields, etc., is practically the same technique as radio echo-sounding. In *impulse radar*, an extremely short pulse of radio frequency radiation is emitted and then analyzed by frequency and time delay to provide a profile through the target material. This technique, in particular, has proved useful in measuring the thickness of sea ice.

2.9   IMAGING RADAR AND SCATTEROMETRY

Imaging radar is a general term to describe active microwave imaging systems. In all of these systems, microwave radiation is beamed toward the Earth's surface from an antenna, and the scattered radiation is collected by the same antenna and used to build up a two-dimensional picture of the backscattering coefficient of the surface. Typical operating frequencies lie between 1 and 10 GHz, although millimeter-wave radars operate at higher frequencies, up to a few hundred GHz. As an active technique, imaging radar is independent of solar illumination. As a microwave technique, it is largely independent of atmospheric propagation effects, although this is less true at the higher frequencies used by millimeter-wave radar, where there is significant continuum absorption by the atmosphere in addition to a number of atmospheric absorption lines.

2.9.1   IMAGING GEOMETRY AND SPATIAL RESOLUTION

In its simplest form, the geometry of imaging radar is illustrated in Figure 2.20. The antenna is long and thin, giving it a "fan beam" region of sensitivity (light shading) that is narrow (angle β) in the direction parallel to the platform motion and broad (angle ψ) in the perpendicular direction. The antenna does not point vertically downward, but somewhat to one side so that the fan beam illuminates the Earth's surface only on this side. The radar transmits very short pulses of microwave radiation from the antenna. We can use the concept of the scattering zone that was introduced in Section 2.7.1 in discussing the radar altimeter. This zone propagates away from the antenna at speed $c/2$, such that any scatterer found within it at time t will contribute to the power received

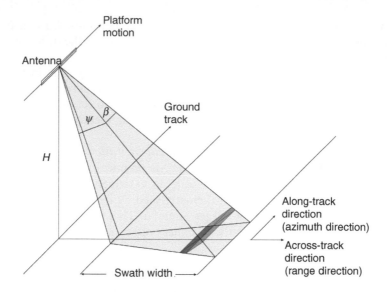

FIGURE 2.20 Geometry of imaging radar (schematic). See the text for an explanation.

by the antenna at that time. The scattering zone is shown in Figure 2.20 in an intermediate shading, and its intersection with the Earth's surface is shown in dark shading. This strip sweeps across the swath, from the near edge to the far edge.

The spatial resolution R_a in the *along-track direction* (also called the *azimuth direction*) is determined by the beam width β. We can put

$$R_a = s\beta$$

where s is the *slant range* from the antenna to the scatterer, and since

$$\beta \approx \frac{\lambda}{L}$$

where L is the length of the antenna and λ is the wavelength of the radiation, and (ignoring the Earth's curvature for simplicity)

$$s = \frac{H}{\cos\theta}$$

where θ is the incidence angle, we see that the azimuth resolution is given by

$$R_a \approx \frac{H\lambda}{L\cos\theta} \qquad (2.26)$$

Thus fine resolution is achieved by having a long antenna and a small value of the height H. The spatial resolution in the *across-track direction* (also called the *range direction*) is determined by the pulse length t_p. Simple geometry shows that the range resolution is given by

$$R_r = \frac{ct_p}{2 \sin \theta}$$

where θ is again the incidence angle. The beam width ψ determines the swath width. Scanning in the across-track direction is achieved by time-resolving the returned signal, somewhat similar to determining the waveform in radar altimetry, while along-track scanning is achieved by the motion of the platform.

The principles that we have just outlined describe the operation of the *real-aperture radar* or *side-looking radar*. Because of the dependence of the azimuth resolution on the height (equation (2.25)), this approach to imaging radar is normally only adopted for airborne systems. It is, however, also used in spaceborne *scatterometry*, in which high spatial resolution is less important than high radiometric resolution. Spaceborne scatterometers usually[*] have several antennas, for example, one looking perpendicular to the platform's motion (as shown in Figure 2.20) together with forward- and aft-viewing antennas at 45° to the direction of motion. This configuration allows the angular dependence of the backscattering coefficient to be determined more comprehensively than from a single antenna.

Let us suppose that we wish to achieve a spatial resolution of 10 m, comparable to what can be obtained from VIR imagers, from a spaceborne imaging radar. If we take nominal values of 5 cm for the wavelength, 800 km for the platform height, and 30° for the incidence angle, equation (2.26) shows that we will require a pulse length of 33 ns or less (which is easy to achieve), but equation (2.25) shows that we will need an antenna that is at least 4.6 km long. This can in fact be achieved, through the technique of *synthetic aperture radar* (SAR). In effect, the data that would have been collected from a very long antenna are synthesized from the data collected as the short antenna is carried along by the platform motion. This requires some complicated signal processing whose details are beyond the scope of this book (see, e.g., Rees (2001) for a discussion of them). Fortunately for the data user, this processing is normally carried out before the data are supplied to the user. In order to process the data in this way, it is necessary that the radiation pulse is transmitted, and the received radiation analyzed, *coherently*, i.e., that both amplitude and phase information are well defined and preserved. This introduces the problem of *speckle*, discussed below.

---

[*]Not all scatterometers use this approach. For example, the SeaWinds instrument carried on the QuikScat satellite uses a conical scan similar to that shown in Figure 2.11.

2.9.2 DISTORTIONS IN RADAR IMAGERY

Side-looking radar and SAR imagery is subject to a number of characteristic geometric and radiometric distortions arising from the oblique viewing geometry and the fact that the across-track (range) coordinate of a scatterer is determined by measuring the time delay of a pulse. This time delay allows the slant range to the scatterer to be determined, but the value of the range coordinate can be determined uniquely from the slant range only if the scatterer's height is known. This point is illustrated in Figure 2.21.

Sometimes radar imagery is supplied using the slant-range coordinate rather than the range coordinate. This is unambiguous but clearly distorted, since the relationship between the slant-range and range coordinates is non-linear. More commonly, the imagery is *geocoded* before being supplied to the user. In this case the relationship between the two ranges has been taken into account. The usual form of geocoding is *ellipsoid geocoding*, in which it is assumed that all scatterers lie on the Earth's ellipsoid. Figure 2.21 serves as a simple model of the errors introduced in this case, by approximating the ellipsoid locally as a horizontal plane. The scatterer at A will be assigned to the position B, equidistant from the radar but on the ellipsoid. The geometry of Figure 2.21 shows that the horizontal displacement between A and B is approximately $H/\tan \theta$, where θ is the incidence angle at A. For example, for

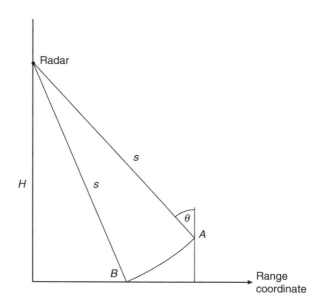

FIGURE 2.21 Scatterers at A and B have the same slant range s from the radar (which is flying "into the page"). If the true position of the scatterer is A but its height coordinate is unknown and assumed to be zero, it will be incorrectly assigned to the position B.

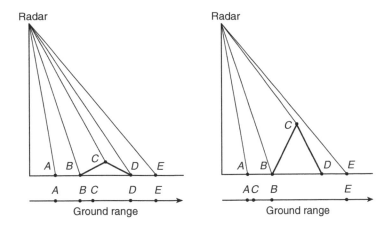

FIGURE 2.22 Layover and shadowing in radar imaging (Rees 2001).

an incidence angle of 23° and a radar height H of 800 km, the terrain-induced displacement due to an elevation of 1000 m would be about 2400 m. (This simple estimate ignores the Earth's curvature.)

Figure 2.22 illustrates two further consequences of the terrain distortion present in ellipsoid-geocoded radar imagery. The figure at the left shows five uniformly spaced scatterers A to E on a simple topography, and the axis labeled "ground range" shows the position of the images of these scatterers in the geocoded image. The hill slope BC that faces the radar has been foreshortened, while the slope CD that faces away from the radar has been stretched out. This is the phenomenon of *layover*. It also introduces a radiometric distortion. If we assume that the terrain is uniformly covered with scatterers between the labeled scatterers A to E, we can see that the power density in the image between B and C will be higher than that between A and B or between D and E, while the power density between C and D will be lower than normal. The brightness of the slope facing toward the radar will thus be enhanced, in a phenomenon called *highlighting*, while the brightness of the slope facing away from the radar will be correspondingly reduced (Figure 2.23). The right-hand side of Figure 2.22 shows a more extreme example of these effects. The slope between B and C has been imaged in reverse order, while the slope between C and D is not imaged at all, since it is not illuminated by the radar. This phenomenon is termed *shadowing*. Unlike optical shadows, radar shadows are completely dark, since the atmosphere does not scatter any radiation into them.

All of these effects, except for shadowing, can be corrected if a sufficiently accurate *digital elevation model* (DEM) of the terrain is available, although the procedure is not straightforward. Imagery in which these effects have been corrected is often described as *terrain-geocoded*. Unlike ellipsoid geocoding, terrain geocoding is not generally offered as an option to the user when obtaining imagery.

FIGURE 2.23 ERS-2 SAR image of part of north-west Svalbard, 21 April 2000. The image is centered on the Brøgger Peninsula (compare Figure 2.8) and covers an area of 100×100 km. The phenomenon of layover is illustrated by the way in which the mountains appear to lean to the right of the image, while highlighting is demonstrated by the tendency of the right-facing slopes to appear much brighter than the left-facing slopes. (Image copyright ESA 2000.)

The distortions that we have just discussed are consequences of the oblique viewing geometry, and the arguments are equally valid for real aperture and synthetic aperture radar. Because it uses the relative motion of the radar and the target to achieve its fine azimuth resolution, synthetic aperture radar is also subject to a number of motion-dependent distortions. The most important of these is *azimuth shift*. If a scatter is moving relative to the Earth's surface, such that the component of its velocity that is directed toward the radar's ground track is u, its image will be displaced in the azimuth (along-track) direction by a distance of approximately

$$\frac{uH \tan \theta}{v}$$

where H is the platform height, θ is the incidence angle, and v is the platform velocity. For a typical spaceborne SAR, the magnitude of this effect can easily be of the order of 50 m per meter per second of u. This is particularly important in observations of the marine environment.

2.9.3 RADIOMETRIC RESOLUTION

The fundamental quantity measured by an imaging radar is the backscattering coefficient σ^0, defined by the *monostatic radar equation*

$$P_r = \frac{\lambda^2 G^2 P_t}{(4\pi)^3 \eta R^4} \sigma^0 A \qquad (2.27)$$

(e.g., Rees 2001). In this equation, P_r is the power received from an area A of the target when it is illuminated by a radar emitting a power P_t. The radar antenna has a gain G and efficiency η and transmits at a wavelength λ, and the distance from the radar to the target is R. σ^0 is a dimensionless variable, essentially the ratio of scattering cross-section to physical area. Because the values of σ^0 observed for natural materials can cover many orders of magnitude, it is usual to specify the value logarithmically, in decibels:

$$\sigma^0(\text{dB}) = 10 \log_{10} \sigma^0 \qquad (2.28)$$

In practice, the variable received from an imaging radar system is normally not σ^0 but the related value β^0, defined in words as the mean radar brightness per unit pixel area and quantitatively through

$$\sigma^0 = \beta^0 \sin \theta \qquad (2.29)$$

where θ is the local incidence angle.

The radiometric resolution of an imaging radar or a scatterometer is usually specified by the uncertainty with which the backscatter coefficient σ^0 can be determined. Since σ^0 is a logarithmic measure of the scattered intensity, this is equivalent to specifying the signal-to-noise ratio of the detected radiation. In the case of coherent imaging systems like SAR, a fundamental limit on the noise is set by the effects of *speckle*, a characteristic granulation in the image that results from interference between the signals received from the many individual scatterers that contribute to a single pixel. In the case of fully developed speckle, the standard deviation in the observed value of σ^0 is large (about 5.6 dB). This can be reduced by averaging, and in fact most radar image products supplied to the user are *multilook* products to which some averaging has already been applied. Scatterometer data are normally incoherently averaged over a large area, so that very high radiometric resolutions (e.g., 0.1 dB or better) are obtained.

2.9.4 SAR INTERFEROMETRY

As we noted in Section 2.9.1, the SAR imaging process requires that both the amplitude and the phase of the scattered signal be retained. Since a phase difference of π radians (half a cycle) corresponds to a change of only a quarter of a wavelength in the one-way distance from the radar to the Earth's surface, and typical wavelengths are only a few centimeters, SAR images potentially contain information of enormously high precision about the geometry of the scattering surface. The technique of SAR interferometry (or *InSAR*) exploits this potential.

The basic idea of SAR interferometry is straightforward. Two SAR images are obtained of the same area of the Earth's surface from slightly different locations. The images are processed in such a way that the phase information is retained (these are usually called *complex images* since they contain both amplitudes and phases, or equivalently, real and imaginary parts). The phase difference between the two images can then be determined. If it can be assumed that this phase difference arises purely from differences between the two imaging geometries, the geometry of the surface can be determined from it, provided that the baseline vector between the two radar locations is known (Figure 2.24).

Perhaps not unexpectedly, the technique is not quite as simple to implement as this outline suggests, and there are many practical complications.

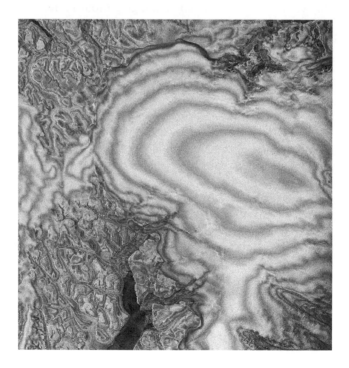

FIGURE 2.24 (See Color Figure 2.5 following page 108) SAR interferogram of Devon Island ice cap, constructed from ERS SAR images acquired on 6 and 7 April 1996. The fringes on the ice cap and elsewhere correspond to changing phase differences between the two images, and thus represent variations in topographic height. (Image provided by courtesy of John Lin, Scott Polar Research Institute.

The usual arrangement for obtaining the two images from space is *repeat orbit interferometry*, in which the satellite returns to almost exactly the same position relative to the Earth's surface after one repeat cycle of the orbit. Since this interval is likely to be several days, it is quite likely that phase differences between the two images may be due to more than just the change in geometry. In this case no interference fringes can be formed, and the technique will fail. The maximum allowable time interval between the two images depends on the dynamics of the surface. An ocean surface can change its geometry on the scale of a few centimeters (the wavelength of the radar radiation) in a matter of seconds, while a glacier surface may be stable at this level for weeks (provided there is no fresh fall of snow), and an exposed rock surface for years.

Further difficulties relate to the *baseline*, i.e., the vector displacement between the locations from which the two images are obtained. If this is too short, the accuracy of the technique is reduced, while if it is too long, the phase variations are so rapid that the SAR images will not be able to resolve them. In practice, the baseline should lie between about 10 m and about 1000 m. It is also necessary that the value of the baseline should be known to an accuracy of better than a wavelength (of the order of a centimeter). In general it is not possible to specify the baseline in advance, so that the acquisition of a pair of images suitable for SAR interferometry is to some extent a matter of chance, and the baseline will not be known to anything like the requisite precision even after the images have been acquired, so that some "fine tuning" is required during the data processing.

Despite these difficulties, SAR interferometry has proved to be a valuable technique for measuring the topography of solid surfaces with accuracies of a few meters or better. It can also be used to measure the bulk displacement of solid surfaces (e.g., glacier flow), with accuracies approaching centimeters.

2.9.5 EXAMPLES

Table 2.4 summarizes the principal spaceborne scatterometers, and Table 2.5 the principal spaceborne SAR systems. Table 2.4 emphasizes the fact that scatterometers provide broad coverage at the expense of low spatial resolution, although the averaging implies that the radiometric resolutions are correspondingly high. Near-global scatterometer data have been continuously available since the launch of ERS-1 in 1991. Table 2.5 omits short-duration SAR operations, such as the SIR (Shuttle Imaging Radar) missions. Again, continuous near-global coverage dates back to the launch of ERS-1. Radarsat (launched 1995) marked a significant development in spaceborne SARs by introducing a multimode system with a diversity of incidence angles, spatial resolutions, and swath widths. The maximum latitudes shown for Radarsat and Envisat are nominal, corresponding to the "standard" imaging mode. The wider-swath modes give greater coverage (Figure 2.25). Radarsat was used for

TABLE 2.4
Principal Spaceborne Scatterometers

Instrument	Satellite	Years	Frequency (GHz) and Polarization	Spatial Res. (km)	Swath Width (km)	Max. Latitude (degrees)
SASS	Seasat	1978	14.6 HH, VV	50	500 × 2	78.3
AMI-Scat	ERS-1, -2	1991–	5.3 VV	50	500	87.8 N, 75.1 S
NSCAT	ADEOS	1996–1997	14.0 HH, VV	50	600 × 2	88.3
SeaWinds	QuikScat	1999–	13.4 HH, VV	50	600 × 2	89.5

TABLE 2.5
Principal Spaceborne Synthetic Aperture Radars

Satellite	Years	Frequency (GHz) and Polarization	Inc. Angle (degrees)	Spatial Res. (m)	Swath Width (km)	Max. Latitude (degrees)	Repeat Period (days)
Seasat	1978	1.3 HH	20	25	100	75.2 N, 68.8 S	N/A
ERS -1, -2	1991–	5.3 VV	23	30	100	84.6 N, 78.3 S	3, 35, 178
JERS-1	1999–1998	1.3 HH	35	18	75	86.2 N, 78.3 S	44
Radarsat	1995–	5.3 HH	20–49	25 × 28	100	88.4 N, 79.1 S	24
			20–31	30–48 × 28	165		
			31–39	32–45 × 28	150		
			37–48	10 × 9	45		
			20–40	50 × 50	305		
			20–49	100 × 100	510		
			50–60	20 × 28	75		
			10–23	28–63 × 28	170		
Envisat	2002–	5.3 HH, VV	15–45	30	56–120	85 N, 78 S	35
				100	400		
				1000	400		

two dedicated "Antarctic Mapping Missions," in 1997 and 2000, respectively (Jezek 2002). The first of these was used to generate the first high-resolution radar mosaic of Antarctica (Jezek 1999) (Figure 2.26). The second had the goal of repeating this mosaic and of measuring the ice velocity field.

2.9.6 SUMMARY

Imaging radars, principally synthetic aperture radars (SARs), are active imaging systems that operate in the microwave region. They can thus obtain

FIGURE 2.25 Wide-swath Envisat SAR image showing part of the Larsen Peninsula and the disintegrating Larsen B ice shelf, 18 March 2002. The image covers an area 400 km wide, and the original data had a spatial resolution of 150 m. (Copyright image reproduced by permission of ESA (http://earth.esa.int/showcase/env/Antarctica/AntarcticPeninsulaLarsenB_ASAR_WS_Orbit00246_20020318.htm)).

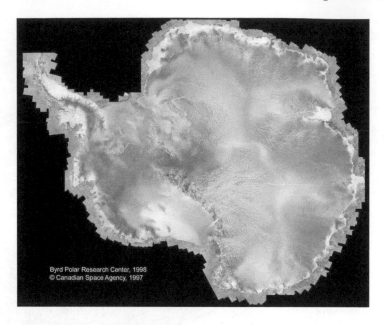

Byrd Polar Research Center, 1998
© Canadian Space Agency, 1997

FIGURE 2.26 Antarctic Mapping Mission mosaic of Antarctica. (Reproduced by courtesy of Byrd Polar Research Center and Canadian Space Agency (http://www.ccrs.nrcan.gc.ca/ccrs/rd/apps/map/amm/amm4_e.html)).

images independently of daylight, and also usually through cloud cover. SARs can provide spatial resolutions comparable to those obtained from VIR imagers, even from space, through sophisticated signal processing. However, these advantages are bought at a price. Imaging radars have an oblique viewing geometry that introduces a number of geometric and radiometric distortions to images of areas showing significant relief, although these can be corrected if an accurate digital elevation model is available. The coherent radiation used for SAR imaging introduced a characteristic "noise", termed speckle, which degrades the radiometric resolution. Scatterometers are variants of imaging radars designed for high radiometric resolution, at the expense of coarser spatial resolution.

SAR images can be combined to obtain extremely precise geometrical data about the Earth's surface, at the level of centimeters, using a technique called SAR interferometry. This technique can be used to measure the topography of solid surfaces to accuracies of the order of a meter, and the bulk translation of solid surfaces to accuracies of the order of a centimeter. It does, however, place considerable technical demands on both the agency operating the SAR and the user processing the data.

3 Image Processing Techniques

3.1 INTRODUCTION

The goal of any kind of remote sensing is to extract useful information from the imagery or other source of data. Sometimes the nature of the required processing is closely linked to the form of the data (this would be true, for example, of radio echo-sounding data). However, in many cases we will be concerned with extracting information from images, and the problems are generic rather than specific. Visual examination of images (both photographs and prints of digital images) has represented an important approach in the past and still has a role to play, but quantitative information is usually extracted by processing digital images in a computer. Digital image processing and image understanding is a huge field of research in its own right, and this chapter can do no more than present a brief outline of the commonest techniques. The reader who wishes to explore the subject in greater depth is referred to the extensive literature on the subject (e.g., the textbooks by Richards 1993; Schowengerdt 1997; Campbell 1996). The treatment in this chapter is similar to the outline of the subject presented by Rees (2001).

Digital image processing can of course be carried out *ab initio* by the user writing suitable computer programs to manipulate the image data. However, there exists a number of software packages, both commercial and noncommercial, that simplify many of the tasks involved. Some of these are combined with *geographic information systems* (GISs), which are essentially computer-based systems for storing and manipulating the kind of information that can be represented on maps.

We begin by defining a digital image.* This is a two-dimensional rectangular array of numerical values,† usually integers, each of which represents the value of the quantity measured by the sensor (typically the radiance of the radiation reaching the sensor from a particular direction). Each cell in this array is a *pixel* (a contraction of "picture element"), and its numerical value is variously referred to as the *gray level*, *digital number*, or *pixel value*. The

---

*Not all remote sensing instruments generate images in the sense in which they are defined here. It may be necessary first to resample the data onto a regular grid.

†This is a definition of a digital image as stored and manipulated by most image processing programs. It is also referred to as *raster format*, in distinction to *vector format* in which the elements of an image are defined in terms of geometrical entities such as points, lines, and polygons. Vector format is a more convenient way of storing the information represented by a map.

coordinates of a pixel within the image can be specified by its *row* and *column* numbers, the row number being measured from the top edge of the image and the column number from the left edge. These numbers usually begin from zero, and they are integers, since they merely count numbers of cells. The region on the Earth's surface that produces the signal represented by a particular pixel is often called a *rezel* (a contraction of "resolution element").* The optics of the sensor are normally arranged so that over a flat, level surface the rezels are rectangular or square. It is common to refer to the size of the rezel as the *spatial resolution* of the instrument, although oversampling may mean that in fact there is some overspill of signal into neighboring pixels in which case the resolution is coarser than the rezel.

The preceding paragraph defines a single-band image, which records the spatial variation of a single quantity. Many sensors record multiband images, in which case the image structure can be thought of as a three-dimensional array, the third dimension corresponding to the spectral band, polarization state, etc.

Before we begin to consider some aspects of digital image processing, it is worth emphasizing the value of just looking at the image, perhaps after some simple preprocessing and image enhancement. The display unit of a computer can present three channels of information, as the intensities of red, green, and blue light. If the image happens to consist of three spectral bands that correspond to the wavelengths of red, green, and blue light it can be displayed as a *true-color composite*. If there are more than three bands, the analyst can choose which three are to be displayed, and how they are distributed among the red, green, and blue channels of the display. Different combinations of bands may be more or less revealing of features of interest.

3.2   PREPROCESSING

Traditionally, the first stages of image processing are referred to as preprocessing, divided into calibration and georeferencing. *Calibration* is the process of converting the raw digital numbers (DNs) measured by the sensor into the physical quantities (e.g., radiances or brightness temperatures) that they represent. In the case of spaceborne imagery this task is normally carried out by the agency operating the instrument, which provides the relevant calibration data to the user. The user may need to go further than this and correct these "at-satellite" values for the effects of propagation through the atmosphere, in order to determine the corresponding "at-surface" values. Procedures for atmospheric correction are beyond the scope of this chapter.

Georeferencing is the process of establishing the relationship between image coordinates (row and column numbers of the pixels) and the corresponding coordinates on the Earth's surface. The latter could be specified as, for

---

*Although it is not strictly logical to do so, it is not uncommon to refer to rezels as pixels.

example, latitude and longitude, or the x and y coordinates of a particular map projection. A common approach to this problem is through the use of *ground control points* (GCPs), which are features that are clearly distinguishable in the image and whose ground coordinates are also known, for example because they have been surveyed. The pairs of coordinates are then used to determine the parameters of a model relating image to ground coordinates. A similar procedure can be adopted to register one image to another without needing to know the corresponding geographical coordinates. In this case, the coordinate system to which the second image is to be referred is that provided by the image coordinates of the first image. An example of this procedure is described in Section 8.8.

The user may wish not only to know the relationship between image coordinates and ground coordinates, but also to *reproject* the image so that the new image coordinates correspond in a simple way to the ground coordinates. For example, it might be desirable that the new row and column directions are aligned with the x and y directions of a map grid. This will in general involve *interpolation* to estimate pixel values for coordinates not represented in the original image. The simplest form of interpolation, which is not really interpolation at all, is called *nearest-neighbor resampling*. As the name implies, it simply consists of choosing the pixel in the old (unprojected) image whose image coordinates are closest to the position required in the new image. More sophisticated techniques such as *bilinear convolution* and *bicubic convolution* estimate the new pixel value as a weighted average of the pixel values in the neighborhood of the corresponding position in the unprojected image. These have the advantage of producing a smoother and more aesthetically pleasing result than nearest-neighbor resampling, but the disadvantage that they generate pixel values that were not measured by the sensor and that may not therefore be physically correct (Figure 3.1). For this reason, nearest-neighbor resampling is generally preferable for quantitative data analysis.

Spaceborne imagery can often be supplied to the user with at least approximate georeferencing information, and reprojected into one of several common map projections.

3.3 IMAGE ENHANCEMENT

Image enhancement is designed to improve the intelligibility of the image as it is seen on the computer monitor or as a hard-copy print. It can be broadly divided into processes that change individual pixel values without reference to neighboring values, and those that take account of spatial context. The former are referred to as contrast modifications, and the latter as spatial filtering. We also include band transformations under this heading. A band transformation is an operation on a multiband image in which the various bands of data, or a subset of them, are recombined.

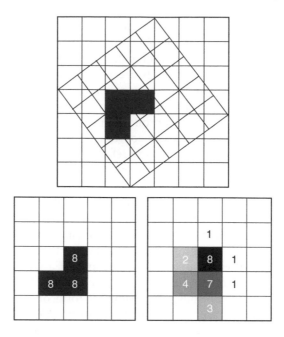

FIGURE 3.1 (Above) Original image is shown as a black grid. Three pixels have values of 8 units, shown as black, while all others have values of zero, shown as white. The image is to be reprojected onto the grid shown in gray. (Below left) The image after reprojection using nearest-neighbor resampling. Pixel values have been preserved but the shape of the feature has been distorted. (Below right) the image after reprojection using bilinear convolution. The shape of the feature is better represented and transitions are smoother, but spurious pixel values have been introduced.

3.3.1 Contrast Modification

The histogram of pixel values present within an image is referred to as the *image histogram*. The effect of a contrast modification is to change the image histogram in some way, by reassigning the pixel values according to a *transfer function* that relates the output pixel values to the input pixel values. The usual reason for doing this is to expand the range of pixel values to make better use of the range that can be displayed by the computer monitor. Most monitors can display a single-band image using 8 bits of data, i.e., the range of integers from 0 to 255. If an image has a histogram that extends over a range of pixel values that is much smaller than 255, it will appear to lack detail. By expanding the range of pixel values used to display it, in an operation called a *contrast stretch*, any detail that is present will be enhanced. An example is shown in Figure 3.2. Here, the transfer function is linear, so the operation is referred to as a *linear contrast stretch*. The fact that the input histogram has a large peak at pixel value 255 suggests that some pixels were saturated in the original image, and of course the process of contrast stretching cannot alter this fact. A more

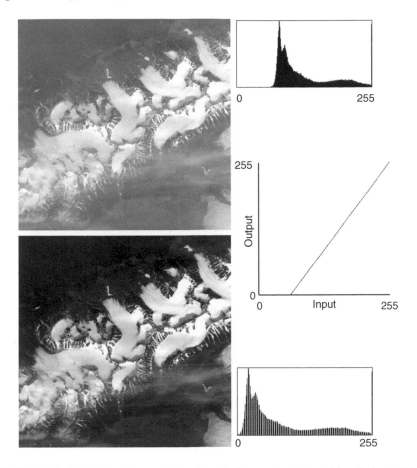

FIGURE 3.2 Left: Original image (above) and after contrast-stretching (below). Right: Corresponding image histograms (above and below) and transfer function (middle). The image is a 12 km square extract of band 1 of a Landsat 7 ETM+ image of the Brøgger Peninsula, Svalbard.

vigorous contrast stretch (one in which the slope of the transfer function is steeper) will increase the number of saturated pixels.

Contrast stretches need not be linear. Other forms of contrast modification can be used to give the output image histogram a specified form, for example a uniform or Gaussian histogram, or one that matches the histogram of another image.

3.3.2 Spatial Filtering

Spatial filtering is applied to images for a variety of reasons. The simplest example is an *averaging* filter, and the most likely reason for applying it

$\frac{1}{9}$	$\frac{1}{9}$	$\frac{1}{9}$
$\frac{1}{9}$	$\frac{1}{9}$	$\frac{1}{9}$
$\frac{1}{9}$	$\frac{1}{9}$	$\frac{1}{9}$

FIGURE 3.3 Convolution kernel for a 3×3 mean filter.

would be to reduce the noise in the image (e.g., the speckle in a SAR image). Suppose we consider a 3×3 pixel moving average in which each pixel value is replaced by the arithmetic mean of the pixel values from the nine pixels within a 3×3 box centered on the pixel. This can be represented schematically by Figure 3.3, which is a representation of the function that is convolved with the original image. For this reason, it can be referred to as a *convolution kernel*. Its interpretation is that this box is centered on a pixel, the nine pixel values covered by the box are multiplied by the respective values in the kernel (the *weights*), these results are added, and the result is the new pixel value to be assigned to the center pixel.

The weights in an averaging filter need not be identical — another popular averaging filter uses a Gaussian function — and the size of the kernel need not be 3×3 pixels, although odd numbers are preferable, since this allows the central pixel to be defined. The fact that the weights in Figure 3.3 sum to one means that the filter does not change the average pixel value of the whole image or of homogeneous regions within it. This is referred to as a *normalized* filter.

Another name for an averaging filter is a *smoothing filter*, since it has the effect of suppressing high spatial frequencies in the image. Conversely, we may wish to enhance high spatial frequencies. Such a filter is usually called a *sharpening filter*, since it makes narrow features such as edges and points more prominent. A typical 3×3 sharpening filter is shown in Figure 3.4. Again, this is normalized.

It is also useful to define the *identity filter*, which leaves the image unchanged. Defined over a 3×3 kernel, this has the form shown in Figure 3.5. While this seems trivial, it allows a useful symbolic notation to be developed. If the identity filter is represented by **I** and an averaging filter is represented by **A**, then any filter

$$\mathbf{S} = (1 + k)\mathbf{I} - k\mathbf{A} \tag{3.1}$$

$\frac{-1}{9}$	$\frac{-1}{9}$	$\frac{-1}{9}$
$\frac{-1}{9}$	$\frac{17}{9}$	$\frac{-1}{9}$
$\frac{-1}{9}$	$\frac{-1}{9}$	$\frac{-1}{9}$

FIGURE 3.4 Convolution kernel for a 3×3 sharpening filter.

0	0	0
0	1	0
0	0	0

FIGURE 3.5 Convolution kernel for a 3×3 identity filter.

is a sharpening filter provided that k, which indicates the degree of sharpening, is greater than zero. For example, the sharpening filter of Figure 3.4 has been generated from the averaging filter of Figure 3.3 by using $k = 1$.

Instead of enhancing the high spatial frequencies (e.g., the sharp edges) in an image while leaving the low spatial frequencies (uniform regions) unchanged, we may want to suppress the low spatial frequencies so that only the high frequencies remain. This is often called an *edge-detection filter* or a *high-pass filter*. Symbolically, a high-pass filter **H** can be generated from an averaging filter **A** by

$$\mathbf{H} = c(\mathbf{A} - \mathbf{I}) \tag{3.2}$$

where c is any constant, positive or negative. Such a filter cannot be normalized, since the weights sum to zero. This means that when the filter is applied it will produce both positive and negative values, so it is usual to add a constant offset to the result. For 8-bit data, this offset is usually chosen to be 128, since this is approximately half way between the lowest integer (0) and highest integer (255) that can be represented.

Another important filter with a total weight of zero is the *gradient filter*. A simple filter that measures the strength of the gradient in the i-direction is shown in Figure 3.6. The corresponding filter for the j-direction is obtained by simply rotating the filter through $90°$.

All of the filters described so far have been linear filters. *Nonlinear filters* are also important. The simplest example is the *median filter*, in which the central pixel of an $n \times n$ region (n is an odd integer) is replaced by the median value. This is particularly useful for filtering SAR speckle, since it is good at preserving edges while suppressing noise. Much more sophisticated nonlinear filters have been developed, generally based on knowledge of the kind of noise that is likely to be present in the image. Examples of the effects of simple spatial filters are shown in Figure 3.7.

0	0	0
$\frac{-1}{2}$	0	$\frac{1}{2}$
0	0	0

FIGURE 3.6 Convolution kernel for a 3×3 gradient filter.

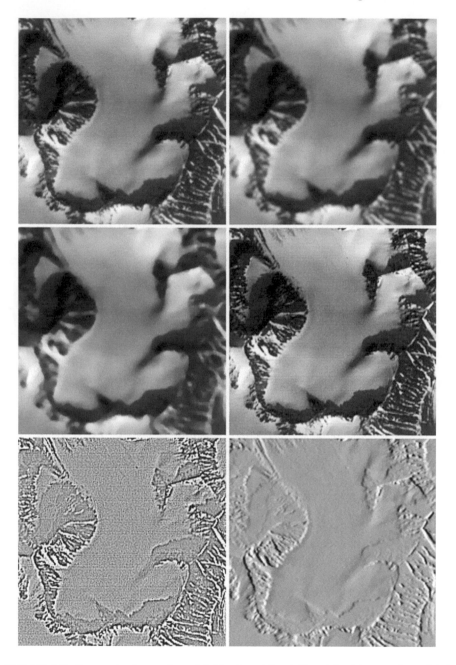

FIGURE 3.7 Demonstrations of spatial filtering. (Top left) original image; (top right) averaging filter; (middle left) median filter; (middle right) sharpening filter; (bottom left) high-pass filter; bottom right: gradient filter. The original image is a 3 km square extract of Band 8 of a Landsat 7 ETM+ image of Midre Lovénbreen, Svalbard.

3.3.3 BAND TRANSFORMATIONS

The image enhancements discussed so far have been applicable to single-band images, or to each band individually in the case of multiband images. Band transformations, by contrast, compare the information present in the different bands of a multiband image.

One of the simplest types of commonly used band transformations forms a ratio image from two bands. This is often useful when the difference between the pixel values in the two bands is diagnostic of some physical quantity or variable to be investigated, but the pixel values also vary because of, for example, variations in the illumination geometry. For example, suppose that a horizontal surface causes a radiance L_1 to be detected by band 1 of a particular sensor, and a radiance L_2 by band 2. If the surface is now tilted toward the direction of the Sun, the band 1 radiance will increase by some factor a, and the band 2 radiance will increase by approximately the same factor. Thus, although the difference between the two bands has increased, the ratio is unchanged. It is shown in Chapter 4 that the reflectance of a deep snow pack is very high (above 0.9) in the wavelength range 0.5 to 0.6 μm, corresponding to band 2 of the Landsat TM or ETM+ sensors, while in the wavelength range 1.55 to 1.75 μm (band 5) it is low (less than 0.2). Thus the ratio of the band 2 reflectance to the band 5 reflectance is expected to be high for snow-covered surfaces (and is in fact diagnostic, as discussed in Chapter 5). In fact, it is preferable to calculate the "normalized difference index" (in this case, the normalized difference snow index or NDSI)

$$\frac{r_2 - r_5}{r_2 + r_5} \tag{3.3}$$

rather than the simple ratio

$$\frac{r_2}{r_5}$$

The reason is that, while the ratio can take any value between zero and infinity and is very sensitive to errors in the denominator if it is close to zero, the normalized difference index (which contains exactly the same information) must lie in the range from −1 to +1. Figure 3.8 shows an example of a normalized difference index.

The second major class of band transformation is the *principal components transformation* (or *principal components analysis*). Suppose we have an n-band image and we represent the pixel value for the i-th band, with given pixel coordinates, by d_i. We can combine these n values into a new pixel value d_1' that is a linear combination of the original values:

$$d_1' = a_{11}d_1 + a_{12}d_2 + \cdots + a_{1n}d_n$$

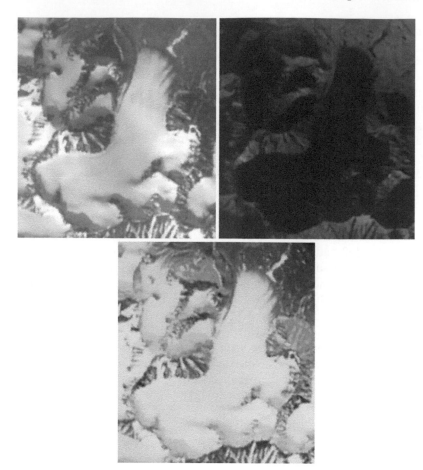

FIGURE 3.8 (Above left) band 2 of a Landsat 7 ETM+ image of Midre Lovénbreen, Svalbard; above right: band 5 of the same image. (Below) the normalized difference snow index (NDSI) calculated from equation (3.3). Note that effects of topographic variation have been greatly reduced.

The coefficients a_{11} to a_{1n} are the same for every pixel in the image. Similarly, we can define a second linear combination

$$d'_2 = a_{21}d_1 + a_{22}d_2 + \cdots + a_{2n}d_n$$

and so on up to[*]

$$d'_n = a_{n1}d_1 + a_{n2}d_2 + \cdots + a_{nn}d_n$$

---

[*]This set of n equations can be more compactly represented as $\mathbf{d'} = \mathbf{ad}$ where $\mathbf{d}$ and $\mathbf{d'}$ are n-component vectors and $\mathbf{a}$ is an $n \times n$ matrix. $\mathbf{d}$ and $\mathbf{d'}$ vary from pixel to pixel but the same matrix $\mathbf{a}$ is used over the whole image.

In a principal components analysis, the set of coefficients a_{11} to a_{nn} is chosen to satisfy the following two conditions:

1. The transformed bands d'_1 to d'_n are uncorrelated with each other.
2. The variance of d'_1 is greater than that of d'_2, which is in turn greater than d'_3, and so on.

The image represented by d'_1 is called the first principal component, and so on. This is a useful transformation because multiband images often show considerable correlation among the bands (one reason for this is the topographic effect discussed earlier), so there is less information than the existence of n bands might imply. An example is shown in Figure 3.9, in which the strong correlation among the six bands of the image can be clearly seen. In this case, the first principal component contains 96.4% of the total variance, and the first three principal components contain 99.8%. The principal components transformation has reordered the information from the original multiband image into new bands of progressively decreasing significance. In this case, it appears that the first three principal components contain virtually all the "interesting" variation present in the original image (and this also means that the original six-band image could be reconstructed almost perfectly from just these three components), while the last principal component contains almost nothing but noise.

The main objection to the principal components transformation is that there is usually no obvious physical interpretation of the transformed bands. However, this is often sufficiently compensated for by the simplification of the image. Principal components analysis need not be confined to multi-spectral imagery; it can also be applied to, for example, multitemporal data to identify spatiotemporal trends. An example of this approach is given in Section 5.2.3.

3.4 IMAGE CLASSIFICATION

Classification is, in essence, the process of turning an image into a map by applying rules to it. The process may be manual, in which the data analyst interprets the image — perhaps after some kind of enhancement — or automated. In practice it is often a combination of both. This section outlines the main approaches to automated image classification, in which rules are applied to the image to determine to which class each pixel belongs.

3.4.1 DENSITY-SLICING

Density-slicing is an operation performed on a single-band image, and it simply consists of identifying all those pixels whose values lie between a certain range (or beyond a certain limit, in which case it can be called *thresholding*). It is thus

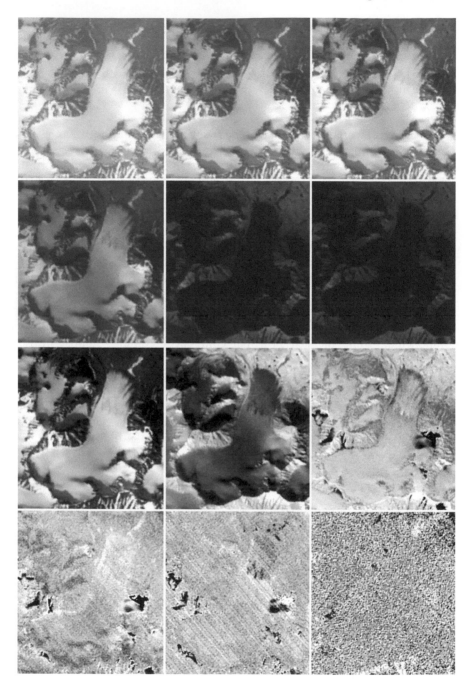

FIGURE 3.9 (See color Figure 3.1 following page 108) (Top) the six reflective bands (1 to 5 and 7) of a Landsat 7 ETM+ image of Midre Lovénbreen, Svalbard. (Bottom) the six principal components of this image.

FIGURE 3.10 NDSI image of Figure 3.8 thresholded at a value of 0.4. The white areas correspond to areas covered by snow.

similar to the idea of making a contour map of the pixel values, and hence makes sense when the pixel values bear a simple relationship to the quantity of interest. Examples of this include thermal infrared and passive microwave data, where the pixel values are related to brightness temperature and hence physical temperature (perhaps the analyst might wish to delineate areas where the physical temperature is above 0°C), and situations where the pixel values corresponding to a class of interest lie at one extreme of the range present in the data. Examples of the latter could include near-infrared imagery, where the reflectance of water is less than that of most other materials, or an image containing an index such as the NDSI (equation (3.3)). By way of example, Figure 3.10 shows a thresholded version of Figure 3.8.

3.4.2 MULTISPECTRAL CLASSIFICATION

Multispectral classification uses the information present in all the bands of a multispectral image to classify each pixel. There are two fundamental concepts in multispectral classification: the idea of a *spectral signature*, in which a particular type of land cover* is represented by a characteristic variation of pixel values across the different bands of the image, and a *classification rule* (or *classifier*) to decide to which of a range of possible land cover types a particular pixel belongs. The classification can be *supervised*, which means that the data analyst defines areas within the image that are known to represent particular land cover types (these are called *training areas*). By performing a statistical analysis of these training areas the corresponding spectral signatures can be established. The classification based on the training areas can thus be thought of as a process of spatial extrapolation — finding pixels that are similar to those representing known land cover types. Alternatively, the classification can be *unsupervised*, in which case the image is scanned automatically to determine the statistically distinct spectral signatures that are present within it. (In some forms of unsupervised classification the user specifies the number of signatures to be found, while in others this number is determined automatically.) In this case, the process of classification can be thought of as identifying all those pixels within the image that are (1) similar to one another, and (2) different from other pixels. The data analyst then has the task of associating these spectral signatures with land cover types.

The two approaches — supervised and unsupervised classification — have complementary characteristics. Supervised classification guarantees that the spectral signatures will correspond to the land cover classes of interest to the data analyst, but on the other hand, it does not guarantee that these signatures will not overlap with one another, causing ambiguity in the classification process. Unsupervised classification is designed to produce distinct, nonoverlapping spectral signatures, but does not guarantee that these will correspond to useful land cover classes (Figure 3.11). In practice, it is common to use a combination of the two approaches. Examples of the use of both approaches are given in Section 8.2.

Another important, though much less commonly used, type of multispectral classification is *spectral mixture modeling*. In this case, the area represented by the image is assumed to consist of a fairly small number of classes and the aim of the classification process is to determine the proportion of each class present within a single pixel. The simplest assumption, and the one that is normally made, is that the spectrum of a pixel (i.e., the variation of pixel values across the bands of the image) is a linear mixture of the spectra of the "pure" classes. For an unambiguous solution, the number of classes cannot exceed the number of bands in the image, so it is a technique suited to simple physical situations (e.g., the image represents only areas of open water and of sea ice) or to images with many spectral bands — so-called *hyperspectral images*. An essential concept in spectral mixture modeling is the *end-member*. End-members are the spectral signatures representing pure image classes.

---

*The term "land cover" is used loosely here to refer to any identifiable class within the image. It need not, of course, be restricted to land surfaces.

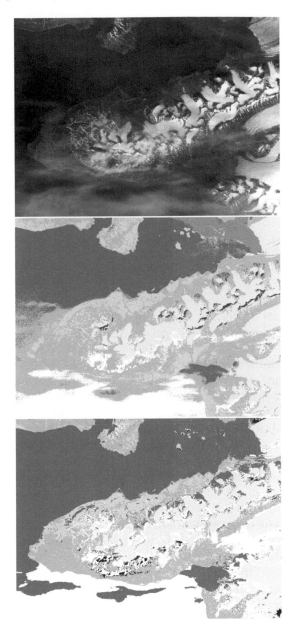

FIGURE 3.11 (See color Figure 3.2 following page 108) (Top) Landsat 7 ETM+ image of Brøgger halvøya, Svalbard. (Middle) Supervised classification (with classes water, water with sediment, bare ground, vegetated ground, snow, shadow, cloud). (Bottom) Unsupervised classification (with classes water, bare ground, vegetated ground, snow, cloud, unclassified). Note the difficulties caused by the presence of cloud in this image; masking the image to remove cloud-covered areas before classification would be advantageous.

They can be determined from field data or (with more difficulty) from the image itself.

Although the discussion in this section has been based on a multispectral image, the methods can be applied to any multiparameter dataset. One obvious extension is to multitemporal data, in which the third dimension of the image (see Section 3.1) represents time rather than wavelength. Another example is given in Section 6.2, where waveform data from a radar altimeter are modeled as linear mixtures of end-members and a procedure similar to spectral mixture modeling is applied.

3.4.3 TEXTURE CLASSIFICATION

The approaches to classification described so far have been *per-pixel* classifiers, in which each pixel is considered in isolation from its neighbors. Texture-based classification, on the other hand, uses the spatial variation within the image as a "feature" on which to base the decision regarding how to classify a pixel. Although the texture-based classification can be implemented pixel by pixel, it is clear that there must be some loss of spatial resolution when compared with per-pixel classifiers. This is because it is necessary to define a region that constitutes the "neighborhood" of each pixel and within which the spatial variation is measured. The larger this region is, the better defined will be the statistics but the poorer will be the spatial resolution, so it is necessary to find a suitable compromise between reliable statistics and higher resolution. Once a texture measure (or a set of measures) has been defined, it can be treated as another variable to describe a pixel, in addition to the pixel value(s). The methods of multispectral classification can then be applied.

Various measures of spatial variation have been adopted for texture-based classification. One of the simplest is just the *standard deviation* (Figure 3.12) or variance of the pixel values within a neighborhood centered on the pixel of interest, or the ratio of the standard deviation to the mean value (the coefficient of variation). Another approach is to measure the *autocorrelation function* of the pixel values and to extract parameters from this function. A third method is based on the *gray-level co-occurrence matrix* (GLCM), also known as the *spatial dependence matrix*. This is a square matrix of size $N \times N$, where N is the number of different pixel values that could be present in the image (so that $N = 256$ for 8-bit data, for example). The element N_{ij} of this matrix is the frequency with which pixel value i occurs at a specified spatial separation from (e.g., one pixel to the left of) pixel value j. Once the GLCM has been calculated, a number of parameters can be derived from it and used as texture measures. The disadvantage of the GLCM is the large number of calculations that are needed. For 8-bit images, the GLCM is defined by $256^2 = 65,536$ elements, although most of these will be zero unless the neighborhood is very large.

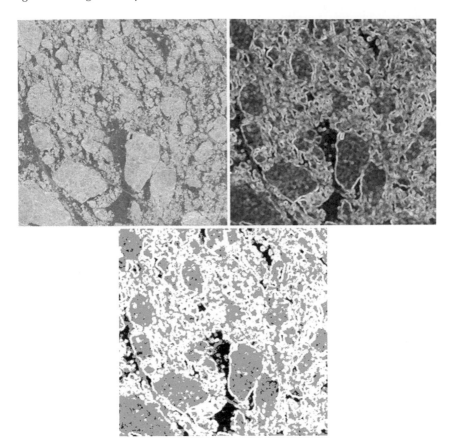

FIGURE 3.12 (Top left) SAR image of sea ice. (Top right) Standard deviation within a two-pixel radius. (Bottom) Density-sliced standard deviation image. The low-texture areas (black) generally correspond to open water and thin ice, while the high-texture areas (white) correspond to floe boundaries and small floes.

Texture-based classification can be particularly useful in cases where multiband data are not available, for example, in single-parameter (single frequency and polarization state) SAR images. It has found application in the characterization of sea ice type (Section 6.3), glacier surfaces (Sections 8.2 and 8.7), and icebergs (Sections 9.2.1 and 9.2.2).

3.4.4 NEURAL NETWORKS

A neural network is, like supervised classification, a procedure that is first trained from known data and then used to classify unknown pixels. It is, however, a much more sophisticated technique than the standard multispectral classification methods. Its name is derived from the fact that it mimics, to a certain extent, the way in which brains learn by reinforcing the links between some neurons. Figure 3.12 illustrates schematically a simple neural network (or *perceptron*). It consists of a number of "layers": an input layer, one or more *processing layers*, also called *hidden layers*, and an output layer. The input layer is supplied with the input data, which might consist, for example, of the pixel values from the several bands of a multiband image, or of texture parameters. The processing and output layers are composed of processing elements, each of which has a number of inputs. The output of a processing element is some function of a weighted sum of the inputs. A typical function is

$$\frac{1}{1 + \exp(-(w_1 x_1 + w_2 x_2 + w_3 x_3 + \theta))} \tag{3.4}$$

where x_1 to x_3 are the inputs (it is assumed that there are three inputs in this example), w_1 to w_3 are the weights (in effect, the strength of the links between processing elements), and θ is an offset. The weights w_1 to w_3 and the offset θ are the parameters of the processing element, and it is the task of the training process to determine them. The example of Figure 3.13 has a single output, which could, for example, be the probability that the pixel belongs to a particular class, but a neural network could have more than one output. Training, also referred to as *back-propagation*, involves using a set of known inputs and corresponding outputs to determine the weights. It is an iterative process.

Although neural network classifiers are substantially more powerful than the simpler classifiers commonly used for multispectral classification, they are more difficult to implement and there is some art involved in designing a suitable network structure. Particular interest has been demonstrated in the use of neural network classifiers for passive microwave data, for example, for the

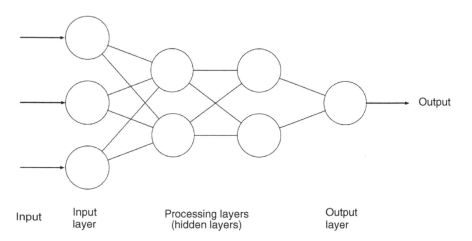

FIGURE 3.13 A simple neural network. See the text for an explanation.

determination of snow water equivalents (Section 5.3.2) and sea ice concentration (Section 6.2).

3.5 DETECTION OF GEOMETRIC FEATURES

We may wish to detect geometric features such as edges, lines, or circles in an image, perhaps to locate the ridges or leads in an area of sea ice (Section 6.6). Edges can be detected by applying a high-pass (edge-detection) filter to the image and then thresholding the result. An example of this is shown in Figure 3.15. Lines of a single pixel in width can be identified using convolution filters, such as those shown in Figure 3.14. These two filters are designed to detect lines running horizontally, and diagonally from bottom left to top right, respectively. Two more filters can be defined by rotating these through 90°. If all four filters are applied to the original image, and the results squared and added, a measure of "line intensity" is obtained, which can then be thresholded. In fact, Figure 3.14 is an example of a more general technique known as *template matching*. If we wish to detect features of a given shape, size, and orientation, it is possible to define a suitable convolution filter, known as a template, which will result in maximum output when it is placed on top of such a feature, and zero output when placed on a uniform area.

Another important class of feature-detection algorithms is the use of *Hough transforms*. The simplest example of a Hough transform is one designed to detect straight lines. A line can be represented by two parameters, essentially its slope and intercept relative to the image coordinate system. Each pixel in the image could lie on any one of a number of lines with different values of these parameters, and this set of possible combinations of parameters is added to an *accumulator array*. After the entire image, or the subset of interest, has been processed, the most likely values of the parameters can be extracted from the accumulator array.

3.6 IMAGE SEGMENTATION

Segmentation is the process of identifying spatially contiguous regions within an image that are homogeneous in some way. For example, one might wish to identify all the icebergs visible in an image, in order to be able to count them

−1	−1	−1
2	2	2
−1	−1	−1

−1	−1	2
−1	2	−1
2	−1	−1

FIGURE 3.14 Templates for line detection.

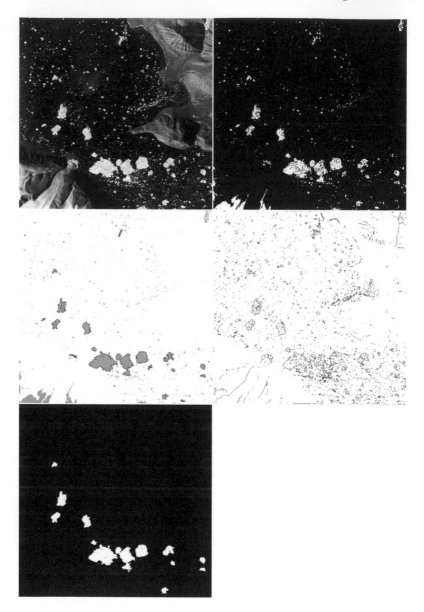

FIGURE 3.15 (Top left) Gray-scale extract of an ASTER image of Dobbin Bay, Ellesemere Island, showing many icebergs. (Top right) original image after thresholding. (Middle left) Segmentation of the thresholded image by boundary-following. Only segments larger than 100 pixels and those that do not touch the edge of the image have been identified, and interior holes have been ignored. (Middle right) Original image after edge-detection followed by thresholding and "skeletonizing," which reduces the width of features to a single pixel. (Bottom left) Results of region-growing from the original image. The seed areas were chosen manually and only the larger icebergs were selected.

and measure their areas. A segment of an image can be defined by its boundary
— the closed curve that surrounds it — so the problem of segmentation is
equivalent to the problem of finding these boundaries. This can be difficult,
and no one method of segmentation is invariably preferable to another.

If the contrast between the objects of interest and the background is large
(e.g., icebergs against a background of open water in a VIR image), simple
thresholding of the image may suffice. This provides a binary classification
of the pixels into "object" pixels and "background" pixels, and it is then a
comparatively simple task to identify contiguous regions of object pixels.
Alternatively, an edge-detection method (Section 3.3.2) can be used to identify
boundaries. If the image is noisy or the contrast between objects and background is not large, the boundaries identified by this procedure can suffer
from a number of defects. They may be discontinuous, and there may be
unwanted edges both within the object and connected to the boundary. A third
approach is *region-growing*. A "seed" pixel or group of pixels is selected within
an object, and each of its neighbors is examined to determine whether it is
sufficiently similar to the object to be added to it. As the region grows by the
addition of pixels, its statistics (and hence the criterion by which new pixels can
be added to it) are modified. Examples of image segmentation are shown in
Figure 3.15, and applications are described in Sections 6.3, 6.6, and 9.2.3.

3.7   CHANGE DETECTION

The last topic in digital image processing that will be briefly considered in this
chapter is the detection of change over time from a series of images. One might,
for example, wish to determine whether a glacier has retreated, and if so by
how much, how sea ice has moved between two dates a few days apart, or
whether the annual variation of snow cover differs significantly at the end
of a decade from 10 years earlier.

Several approaches to change detection are possible. The simplest is direct
comparison of two images, by subtraction or calculating the ratio. This is most
likely to be useful in cases where there is a large and consistent difference
between the pixel values corresponding to the feature of interest and those
corresponding to the background (see Figure 3.16), so that radiometrically
calibrated data, preferably corrected for the effects of solar elevation, are
desirable. Where these conditions are not met, a comparison of classified
images can be made instead. Provided the images are geometrically registered
to each other, it is a simple matter to find regions of the two images where
the image class has changed and those where it has not changed. A similar
approach can be taken with *feature-tracking*, where geometric features are
identified in two (or more) images, the corresponding features in different
images are found, and their displacements hence deduced. This procedure
is often carried out manually, although there is some scope for automating it.
For example, two images of sea ice can be segmented to identify individual

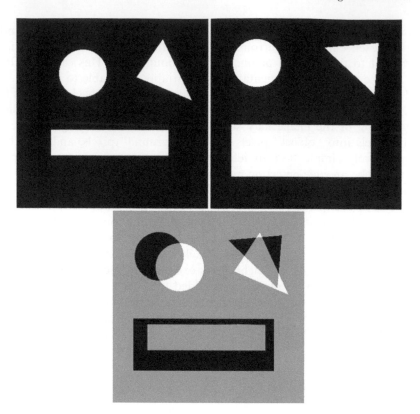

FIGURE 3.16 Schematic demonstration of change detection by image differencing. The difference between the two top images is shown in the bottom image. Gray indicates areas of no change, black shows areas where the white features of the top left image have advanced, and white shows where they have retreated.

floes. If each of these floes is characterized by some parameters that do not depend on shape or orientation, these parameters can be used to match the floes in the two images. A rather simpler problem is to measure relative motion between two images where the features of interest have undergone translation but not rotation. Examples of this could be the migration down a glacier of a pattern of surface features, or the bulk motion of a region of ice floes where the concentration is high enough to prevent relative motion. In such cases the displacement can be determined by calculating the *correlation function* of the two images (or subregions of the images), which will have a maximum value at a position corresponding to the relative displacement. A simple way of calculating the correlation function is through the use of Fourier transforms. An example is shown in Figure 3.17.

Another approach to change detection using two or more (geometrically registered) images is to combine the images into a single image structure so that,

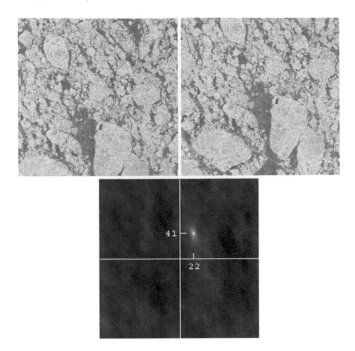

FIGURE 3.17 (Top) Two SAR images; (bottom) correlation function of the two images. The location of the peak in this function shows that the relative displacement between the two images is 22 pixels in the *i*-direction and 42 pixels in the *j*-direction. (The example is artificial, since the two images have been generated from the same original image. This accounts for the unusually prominent peak in the correlation function.)

for example, three monospectral radar images are regarded as being three "bands" of a multispectral image. This combined image can then be analyzed, perhaps by a principal components transformation. The general term for this approach is *composite analysis*. Finally, a brief mention will be made of *change vector analysis*. This can be illustrated by a specific, if rather simplified, example. Suppose we wish to study the long-term variation of the annual pattern of snow cover, we have a technique for estimating the percentage snow cover in each pixel of an image, and we have images for each month of the year. Each pixel can thus be represented by a set of 12 numbers representing the snow cover in each month of the year, and these numbers can be thought of as constituting a (12-dimensional) vector for the year. The long-term, interannual, variation of the snow cover at a particular location is represented by changes in this vector. Analysis of the vectors for the whole image and over a number of years shows the mean trend and the degree of variability, so that regions that behave atypically can be identified.

4 Physical Properties of Snow and Ice

4.1 INTRODUCTION

Remote sensing, as we have already discussed and in the terms in which we have defined it, involves making inferences about the nature of the Earth's surface from the characteristics of the electromagnetic radiation received at the sensor. This process of inference requires that we should establish the relationship between the characteristics of the radiation and the relevant physical properties of the material. These properties and relationships are discussed in this chapter. The "raw materials" of the cryosphere are water and ice, and it would be possible to begin this chapter by considering the chemistry and physics of these materials. Instead, we will introduce ice and snow in their physical contexts rather than as pure examples of the basic materials.

4.2 SNOW

4.2.1 PHYSICAL CHARACTERIZATION

Snow is, in general, a mixture of ice crystals, liquid water, and air. The ice crystals are deposited on the Earth's surface as a result of atmospheric precipitation or wind or mechanical deposition. If the surface of a snow cover has been exposed to a dirty atmosphere for some time, it may acquire a covering of soot or dust.

The most basic intensive physical parameter that describes a snow pack is its *density* ρ_s. Densities lie typically in the range 0.2–0.6 $Mg\,m^{-3}$ unless the snow has fallen very recently in very cold conditions, when it can be as low as 0.01 $Mg\,m^{-3}$ (Rott 1984; Hallikainen, Ulaby, and Van Deventer 1987; Rott, Davis, and Dozier 1992; Guneriussen, Johnsen, and Sand 1996). A typical value for freshly fallen snow (*fresh snow* or *new snow*) is about 0.1 $Mg\,m^{-3}$. As the snow pack ages, its density increases as a result of compaction by wind and gravity, and through thermal metamorphism. A crude empirical model of this process is given by equation (4.1) (Martinec 1977),

$$\rho(t) = \rho_0(1 + t)^{0.3} \tag{4.1}$$

99

where t is the elapsed time in days and $\rho_0 = 0.1 \, \mathrm{Mg \, m^{-3}}$. This implies that snow will reach a density of about $0.3 \, \mathrm{Mg \, m^{-3}}$ after a month and $0.6 \, \mathrm{Mg \, m^{-3}}$ after a year.

The most important parameter describing the internal structure of a snow pack is the *grain size* (or crystal size), often simply defined as the mean radius or equivalent radius of the ice crystals, although some characterizations also take into account the form and orientation of the crystals. Typical grain sizes range between 0.1 and 3 mm, although values as low as 0.01 mm have been reported in new low-density snow. A snow pack can also exhibit larger-scale inhomogeneities, in the form of density variations and macroscopic inclusions of solid ice formed by refreezing of melted snow. These are difficult to characterize, and difficult to model in terms of the interaction of electromagnetic radiation with them.

If the snow pack is below 0°C it is unlikely to contain any liquid water. This state is termed *dry snow*. However, at temperatures at or above 0°C significant quantities of liquid water may also be present. In this case, the snow *wetness* (moisture content, or liquid water content) w is defined as the proportion, usually by volume but sometimes by mass, of the snow pack that is in the form of liquid water. Wetness by volume typically ranges up to about 0.1 (10%). A unit volume of wet snow contains a mass $w\rho_w$ of liquid water, where ρ_w is the density of liquid water and the wetness w is specified by volume, and hence a mass $\rho_s - w\rho_w$ of ice. If the mean volume occupied by an ice crystal is v, the mean *number density* n of the ice crystals in the snow pack is therefore given by

$$n = \frac{\rho_s - w\rho_w}{\rho_i v} \tag{4.2}$$

where ρ_i is the density of ice. The volume v can be estimated on the assumption that the ice crystals are spheres of radius r as

$$v = \frac{4}{3}\pi r^3 \tag{4.3}$$

Taking ρ_w as $1.00 \, \mathrm{Mg \, m^{-3}}$ and ρ_i as $0.92 \, \mathrm{Mg \, m^{-3}}$, and inserting typical values of $0.3 \, \mathrm{Mg \, m^{-3}}$ for ρ_s and 0.5 mm for r, we see that the number density in a dry snow pack is typically of the order of 10^9 crystals per cubic meter.

Related to the density is the snow's *porosity p*. This is defined as the fraction of the total volume that is occupied by air, and determines the extent to which air can diffuse through the snow pack. For wet snow, it follows that

$$p = 1 - \frac{\rho_s - w(\rho_w - \rho_i)}{\rho_i} \tag{4.4}$$

which can be written numerically as

$$p \approx 1 - 1.091\rho_s - 0.091w$$

if the snow density is expressed in Mg m^{-3}. Wet snow is described as *pendular* if the liquid water occupies discontinuous regions and *funicular* if it forms connected paths. The transition from pendular to funicular states occurs when w is around 7%.

The total amount of water contained in a snow pack is specified by the *snow water equivalent* (SWE) d_w. This is defined as the depth of the layer of liquid water that would be produced if all the ice in the snow pack were melted, and is therefore a measure of the mass of water contained per unit area of the snow pack. It is given by

$$d_w = \frac{1}{\rho_w} \int_0^d \rho_s \, dz \tag{4.5}$$

where the integral is carried out vertically through the entire depth d of the snow pack. If the density is uniform, this equation simplifies to

$$d_w = \frac{\rho_s}{\rho_w} d \tag{4.6}$$

so we can see that the SWE will typically be around a third of the depth, although it can be very much lower than this.

4.2.2 SURFACE GEOMETRY

The surface geometry of a snow pack can play an important role in the interaction of electromagnetic radiation with the pack.* Geometric properties can be referred to as *surface roughness* at small length scales and as *surface topography* at larger scales. The distinction depends on the spatial resolution of the application, so in practice it is important to specify the range of length scales over which the property is defined.

The simplest measure of surface roughness is the *root mean square (RMS) height deviation* σ, sometimes imprecisely and perhaps confusingly referred to as the RMS height. This is defined through

$$\sigma^2 = \left\langle \left(h(x,y) - \langle h(x,y) \rangle \right)^2 \right\rangle \equiv \langle (h(x,y))^2 \rangle - \langle h(x,y) \rangle^2 \tag{4.7}$$

In this equation, $h(x, y)$ is the height of the snow surface above some arbitrary level at the position specified by the coordinates (x, y), and the angle brackets $\langle \rangle$ indicate an average over a suitable range of values of x and y. Typical values of σ, measured over a spatial extent of a few tens of centimeters to a meter or so,

---

*This is true not just of snow but of all the materials we will consider, so a full discussion is given here.

range from 0.5 to 30 mm (Hallikainen, Ulaby, and Van Deventer 1987; Rott 1984; Rott, Davis, and Dozier 1992; Rott and Nagler 1992; and author's own unpublished research).

The RMS height deviation σ tells us how much variation in surface height is present, but not (apart from knowing the range of values of x and y over which it was derived) how these variations are distributed spatially. The commonest method of specifying the horizontal scale of the variations is through the use of the *autocorrelation function* (ACF). To save introducing a more complicated notation than is necessary, we first define the ACF for a one-dimensional transect through the snow surface. It is given by

$$\rho(\xi) = \frac{\langle (h(x + \xi) - \langle h(x) \rangle)(h(x) - \langle h(x) \rangle) \rangle}{\sigma^2} \qquad (4.8)$$

In this equation, $h(x)$ is the height transect and σ is the RMS height deviation of the transect. As before, the angle brackets denote spatial averaging.

The ACF, which has no units, is a measure of how similar the profile is to itself when displaced horizontally through a distance ξ. A value of $+1$ shows that the profile is identical to itself; a value of -1 shows that the effect of displacing the profile through distance ξ is to exactly invert the profile, i.e., to replace heights above the mean height by equal heights below the mean height. A value of zero indicates a total lack of correlation. Clearly the ACF is $+1$ when ξ is zero (i.e., $\rho(0) = 1$). A simple way of characterizing the ACF is to specify the *correlation length* l_c, which is defined as the value of ξ for which the ACF first falls to some standard value, such as $1/2$ or $1/e$ (or even zero). Thus, for example, $\rho(l_c) = 1/2$. The correlation length thus becomes a measure of the horizontal scale of the variations in the surface height. Typical values of the correlation length range between 30 and 300 mm (Rott 1984; Rott, Davis, and Dozier 1992); Rott and Nagler 1992; and author's unpublished research).

The ACF is often modeled by a simple function. The commonest of these are the *negative exponential* function

$$\rho(\xi) = \exp\left(-\frac{\xi}{l_c}\right) \qquad (4.9)$$

and the *Gaussian* function

$$\rho(\xi) = \exp\left(-\left(\frac{\xi}{l_c}\right)^2\right) \qquad (4.10)$$

Note that both of these functions have been specified in terms of correlation lengths such that $\rho(l_c) = 1/e$.

Although equation (4.8) defines the one-dimensional ACF, it can be extended in an obvious way to two dimensions. Equivalently, the one-dimensional ACF can be measured in different orientations over the snow surface. If these ACFs are significantly different (if the correlation length is different in different orientations), the surface roughness is said to be *anisotropic*.

We can also define the *slope* of the surface, and derive roughness statistics from it. If the surface height is specified as a function of the two horizontal coordinates x and y, the surface slope is a two-component vector $\mathbf{m}$:

$$\mathbf{m} = \left(\frac{\partial}{\partial x} h(x, y), \frac{\partial}{\partial y} h(x, y) \right) \qquad (4.11)$$

(Sometimes the magnitude and direction of this vector are specified, as slope and *aspect*, instead of its x- and y-components as we have done here.) If we consider just the x-component m_x of this slope, we can define the RMS slope variation σ_{mx} through

$$\sigma_{mx}^2 = \langle (m_x - \langle m_x \rangle)^2 \rangle \equiv \langle m_x^2 \rangle - \langle m_x \rangle^2 \qquad (4.12)$$

and similarly for the y-component. For an isotropically rough surface,

$$\sigma_{mx} = \sigma_{my} = \sigma_m = \sigma \sqrt{ -\frac{d^2}{d\xi^2} \rho(0) } \qquad (4.13)$$

so, for a surface with a Gaussian ACF (equation (4.10)) we see that

$$\sigma_m = \sqrt{2} \frac{\sigma}{l_c} \qquad (4.14)$$

This is an intuitively reasonable result, since it shows that a "typical" surface slope is of the order of the ratio of the typical vertical scale to the typical horizontal scale of the height variations.

It should be noted that not all of the quantities we have just discussed can be meaningfully defined for all surfaces. For surfaces having a negative-exponential autocorrelation function (equation (4.9)) the RMS slope variation cannot be defined, and for some surfaces displaying *fractal* properties the autocorrelation function itself cannot be defined. In this case, the *semivariogram* can be used to characterize the surface properties. In the one-dimensional version, this is defined by

$$\gamma(\xi) = \frac{1}{2} \langle (h(x + \xi) - h(x))^2 \rangle \qquad (4.15)$$

Unlike the ACF, the semivariogram has units and it contains information about both the horizontal and vertical scale of the surface irregularities. Where a meaningful ACF can be defined, it is related to the semivariogram through

$$\gamma(\xi) = \sigma^2 (1 - \rho(\xi)) \qquad (4.16)$$

In one of the simplest kinds of fractal surfaces, for which no meaningful ACF can be defined, the semivariogram has the form of a power law

$$\gamma(\xi) = a\,\xi^{4-2D} \tag{4.17}$$

where a is a constant and D is the fractal dimension of a transect of the surface. D has a value between 1 and 2.

4.2.3 THERMAL PROPERTIES OF SNOW

The thermal properties of snow are important in considering the phenomena of melting and runoff, and the effect that the presence of a snow pack has on heat transfer between the atmosphere and the surface below the snow. The specific latent heat of fusion of ice is $334\,\mathrm{kJ\,kg^{-1}}$, implying that a heat input of $3.34\,\mathrm{MJ}$ would be needed to melt 1 cm of water from one square meter of ice at $0°\mathrm{C}$. The *thermal quality* of a snow pack is defined as the ratio of the heat input required to melt it, relative to that required to melt the same quantity of water from pure ice at $0°\mathrm{C}$. The thermal quality is reduced by the presence of liquid water within the snow pack, and increased by temperatures below $0°\mathrm{C}$. For snow with a thermal quality of 1, the depth of water melted by a rainfall of depth d_r is given by

$$\frac{d_r T_r c_w}{L} \tag{4.18}$$

where T_r is the temperature of the rain above $0°\mathrm{C}$, c_w is the specific heat capacity of water and L is the latent heat of fusion of ice. For values of T_r expressed in $°\mathrm{C}$, this formula becomes approximately (Hall and Martinec 1985)

$$\frac{d_r T_r}{80}$$

In considering the melting of snow by thaw rather than rainfall, the usual approach is to relate the melting, measured as the number of centimeters of water, to the accumulated number of degree–days of thaw, through a coefficient α that has a typical value of $0.5\,\mathrm{cm}$ per degree–day but can be more accurately estimated through

$$\alpha = f(1 - r)\cos\theta \tag{4.19}$$

where r is the surface albedo, θ is the solar zenith angle, and f is a constant. A value of $1.25\,\mathrm{cm}$ per degree–day has been proposed for f (Nagler and Rott 1997).

The thermal conductivity of snow is roughly proportional to its density. At $0.1\,\mathrm{Mg\,m^{-3}}$ it is typically $0.05\,\mathrm{W\,m^{-1}\,K^{-1}}$, similar to fiberglass insulation,

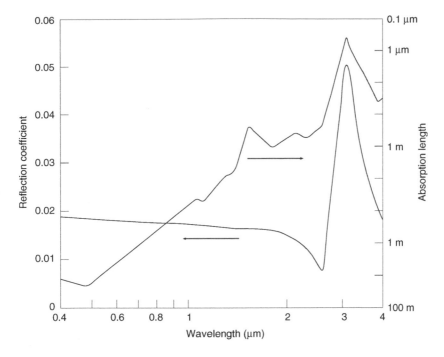

FIGURE 4.1 Reflection coefficient from an ice–air interface and absorption length in ice for electromagnetic radiation with wavelengths between 0.4 and 4.0 μm (simplified).

at $0.3\,\mathrm{Mg\,m^{-3}}$ it is typically $0.13\,\mathrm{W\,m^{-1}\,K^{-1}}$, and at $0.5\,\mathrm{Mg\,m^{-3}}$ it is around $0.44\,\mathrm{W\,m^{-1}\,K^{-1}}$, similar to brick.

4.2.4 ELECTROMAGNETIC PROPERTIES OF SNOW IN THE OPTICAL AND NEAR-INFRARED REGIONS

Fresh dry snow appears white to the human eye. That is to say, it is highly reflective, with little variation over the range of wavelengths (approximately 0.4 to 0.65 μm) to which the eye is sensitive. The reason for this lies in the dielectric properties of ice, and the fact that the ice constituting snow is in a very highly divided form with, as we have seen, of the order of 10^9 particles per cubic meter.

Figure 4.1, which is based on data given by Warren (1984) and collated by Rees (1999), shows why this is so. The absorption length* of visible-wavelength radiation in ice is of the order of 10 m, which means that in traversing a snow

---

*The absorption length is defined as the distance through which the radiation must travel in order for its intensity to be reduced by a factor of e as a result of absorption alone. It can be compared with the *attenuation length*, defined in Section 4.2.5.

pack with a thickness of (say) 2 m, containing a total thickness of perhaps 1 m of ice, a photon has a negligible chance of being absorbed. On the other hand, the photon will encounter of a few thousand air–ice and ice–air interfaces as it traverses the pack, with a probability of about 0.02 of being reflected at each of these interfaces. Thus, it is almost certain that the photon will be scattered back out of the snow pack, and since neither the absorption nor reflection properties of ice vary significantly over the visible waveband, this will be equally true for all wavelengths.

This simple argument, which is developed more fully by Rees (2001), also implies that the reflection coefficient of a snow pack should be smaller if the grain size is larger, since the number of air–ice interfaces and hence scattering opportunities will be reduced. Furthermore, the generally increasing absorption (smaller absorption lengths) at longer wavelengths, illustrated in Figure 4.1, implies a corresponding reduction in reflectance at these wavelengths. These phenomena are illustrated in Figure 4.2, which shows the

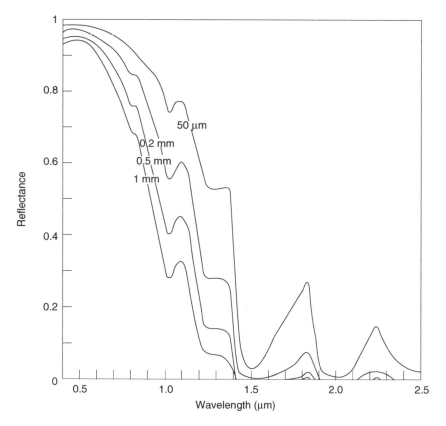

FIGURE 4.2 Spectral reflectance of a deep snow pack as a function of grain size (based on Choudhury and Chang 1979).

comparative insensitivity of reflectance to grain size in the visible region and the high degree of sensitivity in the range from about 1.0 to 1.3 μm.

The reflectance of snow does not depend directly on density, although the processes that cause the increase in density over time represented by equation (4.1) also lead to an increase in grain size and hence a consequent decrease in reflectance. As a snow pack ages, it may also acquire a covering of dust or soot that may also decrease the reflectance. While the albedo of a fresh snow cover can exceed 90%, this figure can fall to 40% or even as low as 20% for dirty snow (Hall and Martinec 1985).

It is clear that these arguments apply only if the snow pack is sufficiently thick, since most of the photons encountering a thin pack will travel right through it without being scattered. This phenomenon can be quantified by specifying a *scattering length* for the snow pack. This is analogous to the absorption length, specifying the distance that radiation must travel through the medium before its intensity in the direction of propagation is reduced by a factor of e as a result of scattering. The *optical thickness* of a uniform snow pack is just the ratio of its actual thickness to its scattering length, and is a measure of its opacity. Since we expect the scattering length to be inversely proportional to both the number density and the cross-sectional area of the grains (see, e.g., Rees 2001), we see from equations (4.2) and (4.3) that it should be proportional to the grain size (radius) and inversely proportional to the snow density. Thus the optical thickness for a given grain size will depend principally on the snow water equivalent. Experimental observations suggest that the SWE that gives an optical thickness of 1 is a few times the grain size.

The presence of liquid water in a snow pack has little direct effect on its reflectance. As was noted in Section 4.2.1, the amount of water rarely exceeds 10% by volume and there is in any case sufficient dielectric contrast between water and ice to ensure that the multiple-scattering phenomenon continues to occur. The absorption of electromagnetic radiation in water is similar to that in ice in the visible and near-infrared regions. On the other hand, the presence of liquid water does have an indirect effect on the optical properties, since it promotes clustering of the ice crystals leading to a larger effective grain size and hence lower reflectance. Green et al. (2002) present model simulations that take into account both grain size and liquid water content as influences on the reflectance of snow. The most significant effect of increasing water content appears to be a small shift of the absorption feature at 1030 nm to shorter wavelengths.

Reflection from a snow pack is anisotropic (not strictly Lambertian*), with enhanced specular scattering due to reflection from ice crystals (Middleton and Mungall 1952; Hall et al. 1993; Knap and Reijmer 1998; Jin and Simpson 1999). However, most studies to date have neglected this effect (König,

---

*A Lambertian surface is one that reflects radiation isotropically, i.e., with equal radiance in all directions, regardless of how it is illuminated. It is an "ideally rough" surface.

Winther, and Isaksson 2001) even though it is rather important in calculating the albedo of a snow-covered surface (see Section 5.5.1).

4.2.5 ELECTROMAGNETIC PROPERTIES OF SNOW IN THE THERMAL INFRARED REGION

Snow is not highly reflective in the thermal infrared region. Figure 4.3 illustrates the variation in reflectance with wavelength from 2 to 14 μm and with grain size from 10 to 400 μm. The figure shows that for grain sizes above 100 μm the reflectance does not exceed 1% throughout the thermal infrared region.

The emissivity of dry snow in the thermal infrared region ranges from typically 0.965 to 0.995. In this region of the electromagnetic spectrum the absorption of ice is high, with a maximum near 10 μm, and the finely divided structure of snow also increases its tendency to act like a black body (i.e., for the emissivity to tend toward 1). Figure 4.4 shows two recent high-resolution measurements of the emissivity spectrum of snow between 3 and 15 μm. Since the emissivity of water in this range of wavelengths is not substantially different from that of snow, the effect of the presence of liquid water is negligible.

A useful summary of the optical and infrared properties of snow is presented by Kuittinen (1997).

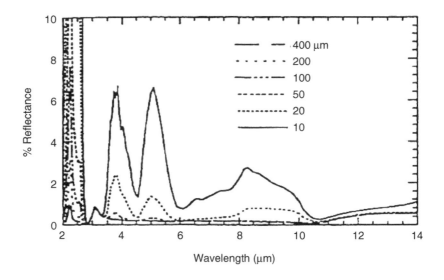

FIGURE 4.3 Spectral reflectance of snow in the thermal infrared region. (Original data from Salisbury, D'Aria, and Wald (1994) reproduced by Kuittinen (1997). Reprinted with permission from European Association of Remote Sensing Laboratories.)

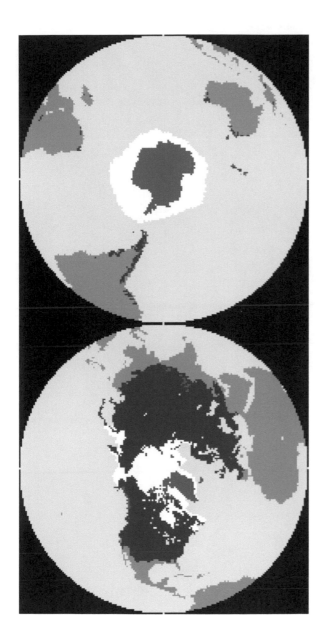

COLOR FIGURE 1.1 Approximate representation of the Earth's snow and ice, at a resolution of 1°. Dark blue shows areas where snow sometimes occurs, light blue shows areas where it is permanent, and white shows areas where sea ice occurs.

COLOR FIGURE 1.2 River ice. (Courtesy of Professor F. Hicks.)

COLOR FIGURE 2.1 Aerial photograph of the part of the accumulation area of Midre Lovénbreen, Svalbard, acquired from an altitude of approximately 1000 m above the glacier surface on 27 July 2004.

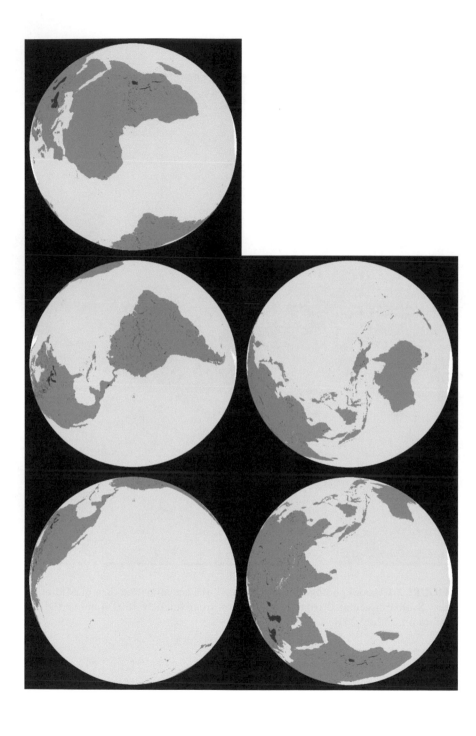

COLOR FIGURE 2.2 Views of the Earth's surface from geostationary satellites. The satellites, with their respective longitudes, are GOES-West (135° W), GOES-East (75° W), Meteosat (0° E), GOMS (76° E), GMS (140° E). Note the limited view of the polar regions and the foreshortening at high latitudes.

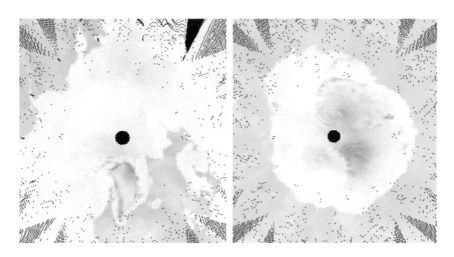

COLOR FIGURE 2.3 Color composites of SSM/I passive microwave radiometer data (red = 19H, green = 22V, blue = 37H). Left: 1 January 2004; right: 1 June 2004. (Original data downloaded from Maslanik and Stroeve [2004].)

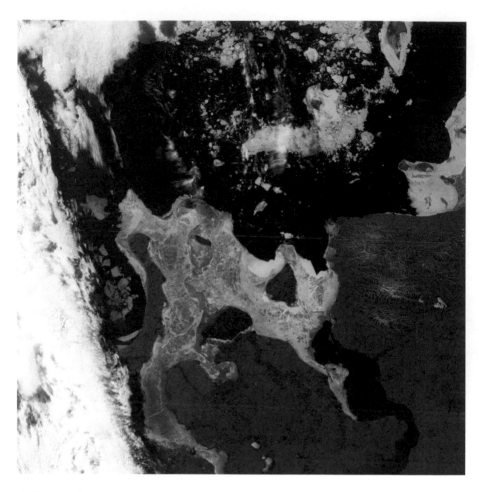

COLOR FIGURE 2.4 Extract of MSU-SK image of the mouth of the Yenisei River, Russia, 10 July 1998. The extract, which covers an area roughly 500 km square, has been presented as a false-color infrared composite, and shows sea ice, icebergs, lake ice, a light snow cover at center right, and a band of cloud at the left side of the image. The original image has a spatial resolution of approximately 150 m.

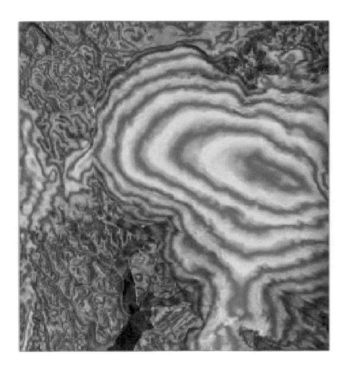

COLOR FIGURE 2.5 SAR interferogram of Devon Island ice cap, constructed from ERS SAR images acquired on 6 and 7 April 1996. The fringes on the ice cap and elsewhere correspond to changing phase differences between the two images, and thus represent variations in topographic height. (Courtesy of John Lin, Scott Polar Research Institute.)

COLOR FIGURE 3.1 The first three principal components of a Landsat-7 ETM+ image of Midre Lovénbreen, Svalbard, presented as a multispectral composite. (Left) red–green–blue composite (PC1 is shown in red, PC2 in green, and PC3 in blue); (right) brightness–saturation–hue composite.

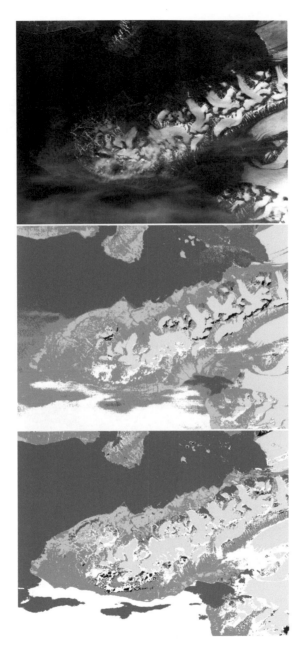

COLOR FIGURE 3.2 (Top) Landsat 7 ETM+ image of Brøgger halvøya, Svalbard. (Middle) Supervised classification (with classes water, water with sediment, bare ground, vegetated ground, snow, shadow, cloud). (Bottom) Unsupervised classification (with classes water, bare ground, vegetated ground, snow, cloud, unclassified). Note the difficulties caused by the presence of cloud in this image; masking the image to remove cloud-covered areas before classification would be advantageous.

COLOR FIGURE 5.1 MODIS true-color image of snow cover over Scandinavia and north-western Russia, 15 March 2002. (Reproduced from the NASA Visible Earth web site at http://visibleearth.nasa.goc.)

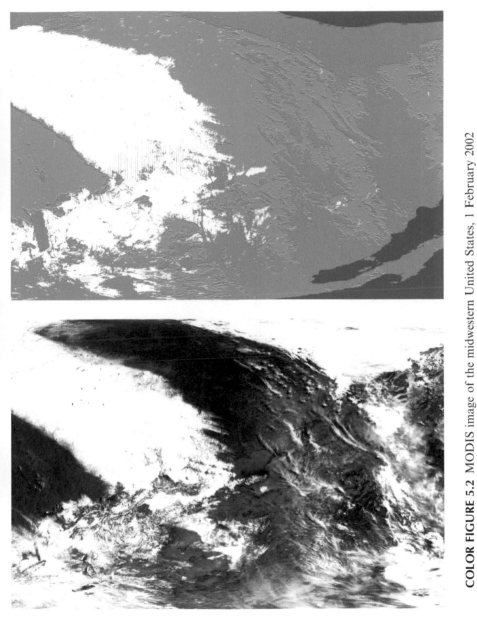

COLOR FIGURE 5.2 MODIS image of the midwestern United States, 1 February 2002 (http://modis-snow-ice.gsfc.nasa.gov/MOD10_L2.html). (Left) multispectral composite; (right) corresponding classified snow-cover image, showing snow (white), cloud (purple), snow-free land (green), and water (blue).

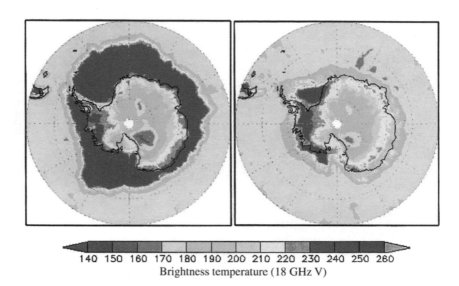

140 150 160 170 180 190 200 210 220 230 240 250 260
Brightness temperature (18 GHz V)

COLOR FIGURE 6.1 Brightness temperature over Antarctica measured with the Oceansat-1 Multichannel Scanning Microwave Radiometer (MSMR). (Left) 12 to 18 September 1999; (right) 13 to 19 March 2000. (Adapted from Dash et al. (2001), with permission of Taylor & Francis plc.)

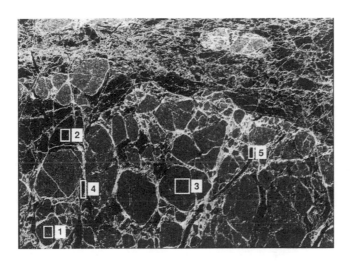

COLOR FIGURE 6.2 Multifrequency airborne SAR image of sea ice in the Beaufort Sea, presented as a false-color composite image. Red: P band; green: L band; blue: C band. (Adapted from Drinkwater et al. (1992), copyright American Geophysical Union 1992.)

COLOR FIGURE 6.3 Interferogram derived from ERS-1 SAR images of the northern end of the Gulf of Bothnia, 27 and 30 March 1992. One cycle of phase corresponds to a movement of 28 mm. The linear features at "2" are the tracks of icebreakers; the incoherent region at bottom center is open water. (Reproduced from Dammert, Leppäranta, and Askne (1998), with permission of Taylor & Francis plc.)

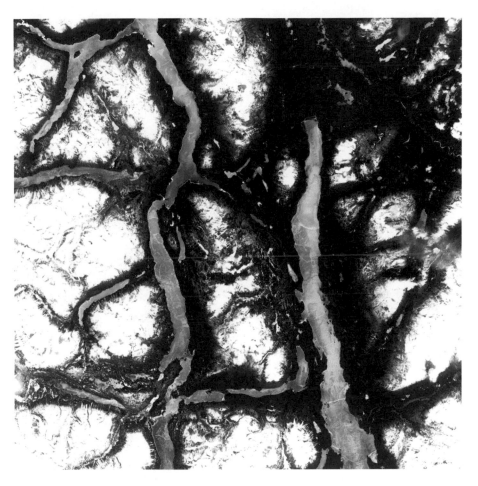

COLOR FIGURE 7.1 Extract of Landsat-7 image of Tagish Lake, British Columbia, 8 May 2000. (Reproduced from the NASA Visible Earth web site http://visibleearth.nasa.gov by courtesy of Brian Montgomery, NASA Goddard Space Flight Center.)

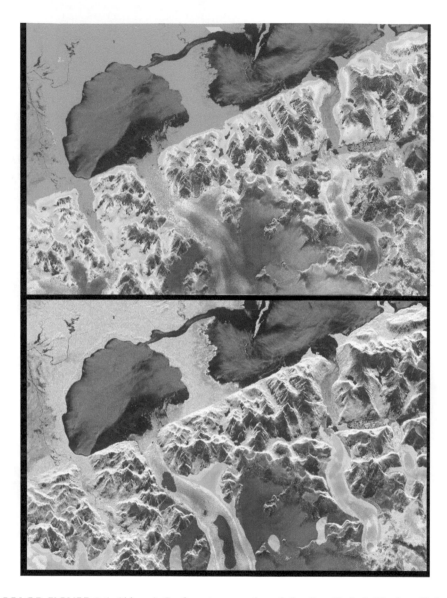

COLOR FIGURE 8.1 (Above) Surface topography of the San Rafael Glacier, Chile, derived from SAR interferometry using L-band data collected by the SIR-C/X-SAR mission between 9 and 11 October 1994. Elevation is represented by color, from blue (sea level) to pink (2000 m), while the brightness of the image represents the backscattering coefficient. (Below) Component of the ice velocity parallel to the look azimuth of the radar, ranging from more than 6 cm/day away from the radar (purple) to more than 180 cm/day toward the radar (red). (Reproduced from the NASA Visible Earth web site http://visibleearth.nasa.gov by courtesy of NASA Jet Propulsion Laboratory.)

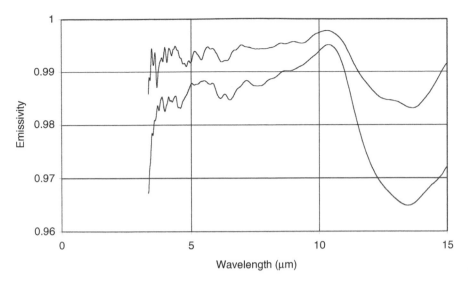

FIGURE 4.4 Thermal infrared emissivity of two snow samples (Zhang 1999).

4.2.6 ELECTROMAGNETIC PROPERTIES OF SNOW IN THE MICROWAVE REGION

The simplest phenomenon to consider is the reflection of radiation from a planar snow surface. This is controlled by the angle at which the radiation strikes the surface, and the dielectric constant of the snow. The greater the difference between the dielectric constant of snow and that of the external medium (air, although without significant error we can assume it to be free space with a dielectric constant of 1), the greater the reflection coefficient.

The real part of the dielectric constant of ice is practically constant throughout the microwave region, with a value of 3.17. Consequently, the real part of the dielectric constant of dry snow depends only on the snow density. It can be expressed numerically as (Hallikainen, Ulaby, and Abdelrazik 1986)

$$\varepsilon' = 1 + 1.9\rho_s \tag{4.20}$$

where the snow density ρ_s is expressed in $Mg\,m^{-3}$. Thus, the real part of the dielectric constant at a typical density of $0.3\,Mg\,m^{-3}$ is around 1.57.

The imaginary part of the dielectric constant, which determines the degree of absorption, is very small for dry snow. It exhibits some dependence on temperature. Figure 4.5 shows the prediction of semiempirical models of the imaginary part of the dielectric constant.

Because of the low values of the absorption coefficient, propagation of microwave radiation through dry snow is generally dominated by scattering (as we saw in Section 4.2.4, this is also true for visible and near-infrared radiation).

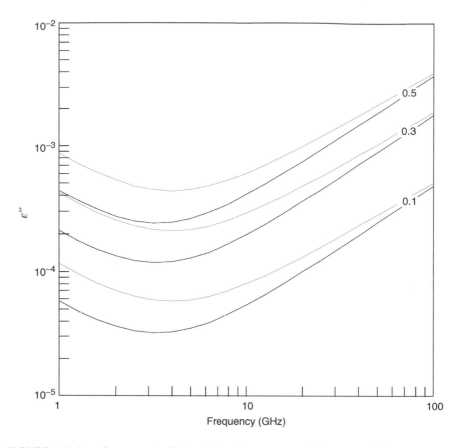

FIGURE 4.5 Imaginary part of the dielectric constant of dry snow. The curves are labeled with the density (in $Mg\,m^{-3}$) and are based on the work of Mätzler and Wegmüller (1987); Tinga, Voss, and Blossey (1973). Gray curves are for a temperature of $-5°C$, black curves for $-15°C$.

Figure 4.6 shows the *attenuation length*[*] for microwave radiation in dry snow. The dependence on grain size is in the opposite sense to what is observed at visible wavelengths. In the latter case, the grain size r is much larger than the wavelength and the scattering cross-section of an individual grain is roughly proportional to r^2. In the case of microwave radiation, the grains are significantly smaller than the wavelength and the scattering cross-section is proportional to a higher power of r, tending to the sixth power in the limit of very small grains (the case of *Rayleigh scattering*).

---

[*]Attenuation length is defined similarly to absorption length. It is the distance through which radiation must travel in order for its intensity to be reduced by a factor of e when both absorption and scattering are taken into account. It is sometimes called the *penetration length* or *penetration depth*.

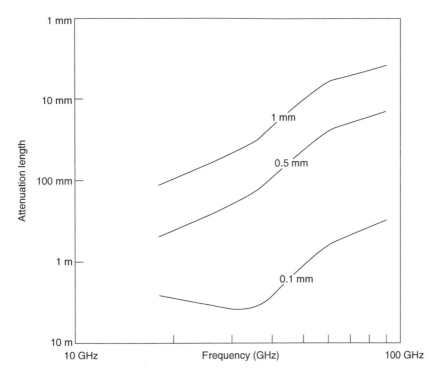

FIGURE 4.6 Empirical attenuation lengths in snow as a function of grain radius and frequency (based on Hallikainen, Ulaby, and Van Deventer 1987).

The microwave dielectric properties of wet snow have been comprehensively reviewed by Hallikainen and Winebrenner (1992), drawing on experimental and theoretical studies performed by various workers during the 1980s. They proposed the following empirical model for frequencies between 3 and 15 GHz:

$$\varepsilon' = 1 + 1.83\rho_{ds} + 0.02w^{1.015} + \frac{0.073w^{1.31}}{1 + (f/9.07)^2} \tag{4.21}$$

$$\varepsilon'' = \frac{0.073(f/9.07)w^{1.31}}{1 + (f/9.07)^2} \tag{4.22}$$

At frequencies between 15 and 37 GHz, the model is somewhat more complicated:

$$\varepsilon' = 1 + 1.83\rho_{ds} + \left(0.016 + 0.0006f - 1.2 \times 10^{-5}f^2\right)w^{1.015}$$

$$+ 0.31 - 0.05f + 8.7 \times 10^{-4}f^2 + \frac{\left(0.057 + 0.0022f - 4.2 \times 10^{-5}f^2\right)w^{1.31}}{1 + (f/9.07)^2}$$

$$\tag{4.23}$$

$$\varepsilon'' = \frac{(0.071 - 2.8 \times 10^{-4}f + 2.8 \times 10^{-5}f^2)(f/9.07)w^{1.31}}{1 + (f/9.07)^2} \tag{4.24}$$

In these expressions, f is the frequency in GHz, w is the percentage liquid water content by volume, and ρ_{ds} is the mass of ice per unit volume of snow, expressed in $\mathrm{Mg\,m^{-3}}$. The implications of these expressions are shown in Figure 4.7. In particular, the figure shows the very strong effect on absorption of only a small amount of liquid water. For example, at 10 GHz the absorption length, which can be of the order of a meter or so for dry snow, falls to only a few centimeters at a liquid water content of just 2%. The figure also shows that the imaginary part of the dielectric constant, and consequently the absorption length, are only weakly dependent on the density of the snow at a given frequency and water content.

A simpler model (Rott, Mätzler, and Strobl 1986) proposes that the dielectric constant for dry snow should be increased by

$$\Delta\varepsilon' = \frac{23w}{1 + (f/f_0)^2} \tag{4.25}$$

and

$$\Delta\varepsilon'' = \frac{23(f/f_0)w}{1 + (f/f_0)^2} \tag{4.26}$$

to account for the presence of liquid water, where w is the volumetric water content (*not* expressed as a percentage), f is the frequency, and $f_0 = 10\,\mathrm{GHz}$. This simpler model tends to overestimate the absorption at low values of the liquid water content.

4.2.7 MICROWAVE BACKSCATTERING FROM SNOW

Scattering of radiation from a snow surface (in fact, from any surface) depends upon the dielectric constant of the surface, its roughness properties, and the geometry of the scattering. Many mathematical models of *surface scattering* have been developed, some based on physical laws, some on empirical data fitting, and some on a combination of the two. The physically based models are usually more complicated than the others but have a wider range of applicability.

The simplest physical model of microwave scattering from a "moderately" rough surface is the *stationary phase model* or *geometric optics model*. This gives the backscattering coefficient as

$$\sigma^0_{HH}(\theta) = \sigma^0_{VV}(\theta) = \frac{|R(0)|^2 \exp\left(-(\tan^2\theta/2\sigma^2_m)\right)}{2\sigma^2_m \cos^4\theta} \tag{4.27}$$

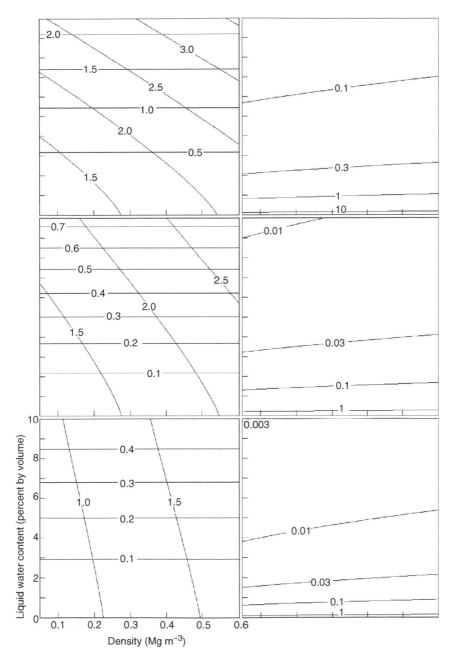

FIGURE 4.7 Calculated dielectric constants (left) and absorption lengths (right) of wet snow at 3 GHz (top), 10 GHz (middle) and 30 GHz (bottom). The plots of dielectric constant are labeled with the real part (sloping lines) and imaginary part (horizontal lines). The plots at the right are labeled with the value of the absorption length in meters.

where θ is the incidence angle and σ_m is the RMS slope variation defined in equation (4.12). $R(0)$ is the amplitude reflection coefficient for radiation incident normally on a planar snow surface, and is given by

$$R(0) = \frac{1 - \sqrt{\varepsilon}}{1 + \sqrt{\varepsilon}} \tag{4.28}$$

where ε is the complex dielectric constant of the surface material. The conditions for validity of this model are usually stated as

$$\frac{\sigma \cos \theta}{\lambda} > 0.25$$

$$\frac{l_c}{\lambda} > 1$$

$$\frac{l_c^2}{\sigma \lambda} > 2.8$$

where λ is the wavelength of the radiation, σ is the RMS height varia-tion (equation (4.7)), and l_c is the correlation length of the surface irregularities (equation (4.8)). For a surface with a Gaussian autocorrelation function (equation (4.10)) these conditions can be rewritten as

$$\frac{\sigma \cos \theta}{\lambda} > 0.25$$

$$\frac{\sigma}{\lambda} > 0.7\sigma_m$$

$$\frac{\sigma}{\lambda} > 1.4\sigma_m^2$$

If the surface is too smooth for the first of these conditions to be satisfied, but $\sigma_m < 0.25$, the somewhat more complicated *scalar approximation* can be used. This gives the backscattering coefficient as

$$\sigma_{pp}^0 = \frac{8\pi^2}{\lambda^2} \cos^2 \theta \left| R_p(\theta) \right|^2 \exp\left(-\frac{16\sigma^2 \cos^2 \theta}{\lambda^2} \right)$$

$$\times \sum_{n=1}^{\infty} \frac{\left(16\sigma^2 \cos^2 \theta / \lambda^2 \right)^n}{n!} \int_0^{\infty} \rho^n(\xi) J_0\left(\frac{4\pi\xi \sin \theta}{\lambda} \right) \xi \, d\xi \tag{4.29}$$

where σ_{pp}^0 is the backscattering coefficient for *pp*-polarized radiation (i.e., *p* is either H or V for horizontally or vertically polarized radiation respectively), $R_p(\theta)$ is the amplitude reflection coefficient for *p*-polarized radiation incident on a plane surface at angle θ, and $\rho(\xi)$ is the surface autocorrelation function (equation (4.8)). If the surface can be described by a Gaussian autocorrelation function, the formula can be slightly simplified to

$$\sigma_{pp}^0 = \frac{4\pi^2 l_c^2}{\lambda^2} \cos^2 \theta \left| R_p(\theta) \right|^2 \exp\left(-\frac{16\sigma^2 \cos^2 \theta}{\lambda^2} \right)$$

$$\times \sum_{n=1}^{\infty} \frac{\left((16\sigma^2 \cos^2 \theta)/(\lambda^2) \right)^n}{n!n} \exp\left(-\frac{4\pi^2 l_c^2 \sin^2 \theta}{n\lambda^2} \right) \tag{4.30}$$

Scattering from within the bulk of a snow pack is particularly important when the snow is dry and the absorption length is correspondingly long. In *volume scattering*, the radiation that passes through the surface is scattered by the irregularities within the bulk of the medium, in this case the ice crystals within the snow pack. The most straightforward model of volume scattering in a snow pack was put forward by Stiles and Ulaby (1980b). In its simplest form, this can be written as

$$\sigma^0(\theta) = \left| R(\theta) \right|^4 \frac{\sigma_v \cos \theta'}{\gamma} \left(1 - \frac{1}{L^2(\theta')} \right) + \left| R(\theta) \right|^4 \frac{\sigma_g^0(\theta')}{L^2(\theta')} \tag{4.31}$$

In this expression, σ_v is the volume scattering coefficient and γ is the attenuation coefficient (i.e., the reciprocal of the attenuation length). Rott et al. (1985) have proposed that the volume scattering coefficient can be simply modeled as half of the scattering coefficient, i.e., half the reciprocal of the scattering length. θ' describes the angle that the radiation makes with the normal to the surface after it has been refracted into the snow pack, and L is the propagation loss factor, given by

$$L(\theta') = \exp\left(\frac{\gamma d}{\cos \theta'} \right) \tag{4.32}$$

where d is the depth of the snow pack. σ_g^0 is the backscattering coefficient of the underlying (ground) surface.

Equations (4.31) and (4.32) show that if the loss factor L is close to 1, the volume scattering will be insignificant compared with the scattering from the underlying surface. It was shown in Figure 4.6 that the attenuation length for microwaves in dry snow can be long — of the order of a meter or so in the case of fine-grained snow — and in consequence, dry snow packs can be effectively almost transparent unless they are deep, giving a backscattering coefficient that is essentially that of the underlying surface.

Figure 4.8 illustrates the typical dependence of the backscattering coefficient with the incidence angle for wet snow. Although it has been constructed for a nominal frequency of 10 GHz, its general form is typical. The total

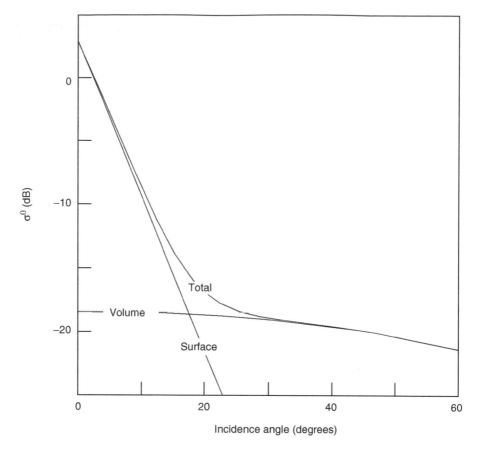

FIGURE 4.8 Typical variation of copolarized backscattering coefficient with incidence angle for wet snow (schematic).

backscatter contains contributions from both surface and volume scattering. The former is strongly peaked near normal incidence, while the latter shows only a very slow variation with incidence angle. As a consequence, surface scattering tends to dominate for incidence angles near zero, and volume scattering tends to dominate at large incidence angles.

A theoretical and experimental investigation of microwave scattering from snow packs, which includes a discussion of polarization effects, is presented by Kendra, Sarabandi, and Ulaby (1998).

4.2.8 MICROWAVE EMISSION FROM SNOW

As has been shown in Section 4.2.7, a dry snow pack is comparatively transparent to microwave radiation, and the attenuation is generally dominated by scattering. Consequently, thermal emission from dry snow originates from a range of depths and hence cannot be simply characterized by a single value of the emissivity. This problem is compounded by the fact that the temperature of a deep snow cover is not uniform with depth, as a result of the time lag associated with the conduction of heat through the snow pack. If the snow pack were uniform, the effective bulk emissivity would depend on the dielectric constant of the ice, the grain size, and the depth. The effect of grain size is highly significant in this context. Larger grains give rise to much smaller volume scattering coefficients and hence to smaller values of the effective emissivity. Theoretical calculations (Chang et al. 1976) for a 20 m snow pack at a physical temperature of 250 K showed that, at a frequency of 10.7 GHz, the surface brightness temperature would be around 250 K for a grain size of 0.2 mm, falling to around 150 K at 1 mm and 10 K at 5 mm. At higher frequencies the decrease in brightness temperature with grain size is shifted to smaller grain sizes: at 19.4 GHz the approximate brightness temperatures for grain sizes of 0.2, 1, and 5 mm are 250, 65, and 10 K respectively, while at 37 GHz these values become 150, 20, and 20 K respectively.

The effect of grain size on the effective emissivity has been demonstrated empirically by Sturm, Grenfell, and Perovich (1993) for tundra and taiga snow. Both of these snow covers are stratified, with a fine-grained surface overlying a layer of coarse-grained depth hoar. At 18 GHz, the authors reported emissivities for the surface snow of around 0.96 to 0.98 (vertical polarization) and 0.93 to 0.98 (horizontal polarization), compared with 0.90 to 0.94 (vertical) and 0.80 to 0.89 (horizontal) for the depth hoar. At 37 GHz the differences were more pronounced: The surface snow showed emissivities of around 0.85 to 0.97 (vertical) and 0.83 to 0.97 (horizontal), while the depth hoar showed values of 0.65 to 0.72 (vertical) and 0.60 to 0.68 (horizontal). All these data were acquired at an incidence angle of 50°.

Data on the effect of incidence angle and polarization have been reported by Rott, Sturm, and Miller (1993), and are summarized in Table 4.1. This shows the calculated value of the effective emissivity, averaged over three sample sites of Antarctic firn. The data show that while the emissivity in horizontal polarization declines monotonically with increasing incidence angle, the vertically polarized emissivity shows a maximum at around 50°. This is essentially a manifestation of the Brewster angle phenomenon.

The effect of even partial melting on the microwave emission from a snow pack is strong. The presence of wet snow decreases the significance of volume emission in favor of surface emission, and increases the effective emissivity (Stiles and Ulaby 1980a; Rango, Chang, and Foster 1979). At a frequency of 37 GHz, the increase in brightness temperature as a result of melting can be as much as 50 K (Hofer and Mätzler 1980). At still higher frequencies the effect of melting on the brightness temperature increases somewhat. The nadir emissivity of snow at frequencies between 24 and 157 GHz as reported by Hewison and English (1999) is between about 0.63 and 0.72 for dry snow and about 0.96 for wet snow.

TABLE 4.1
Typical Effective (Bulk) Emissivities of Antarctic Firn

Incidence Angle (degrees)	5.2 GHz		10.3 GHz	
	V	H	V	H
10	0.88	0.85	0.84	0.82
20	0.87	0.85	0.84	0.81
30	0.90	0.83	0.85	0.80
40	0.91	0.81	0.87	0.78
50	0.91	0.78	0.88	0.75
60	0.91	0.75	0.88	0.71
70	0.87	0.70	0.85	0.66
80	0.71	0.56	0.71	0.54

After Rott, Sturm, and Miller (1993).

4.3 LAKE AND RIVER ICE

4.3.1 PHYSICAL CHARACTERIZATION

Lake and river waters are typically not saline, with a total content of dissolved solids usually less than 0.2 parts per thousand (compared with typically 34 parts per thousand in the ocean). This is insufficient to cause any significant lowering of the freezing point, and lake and river ice can be considered to be "pure" ice, freezing at 0°C (Welch 1991).

The initial stage of lake freezing is the formation of millimeter-sized platelet and needle-shaped crystals, termed *frazil*, near the surface. (A similar process occurs in the formation of sea ice, where the term frazil is also used.) Although these crystals are initially randomly oriented, preferential growth along the crystallographic *c*-axis means that once they have formed a continuous layer about 10 cm thick the lower surface consists entirely of crystals with the *c*-axis horizontal. This form of ice is termed *congelation ice* or *black ice*, and it is practically pure freshwater ice in the form of a smooth transparent sheet with few or no inclusions or surface irregularities. (Congelation ice also refers to a form of sea ice, described in Section 4.4.1.) If a lake ice surface is heavily loaded with snow, however, the weight can depress the surface, and water can rise through any cracks in the ice, saturating the overlying snow. This layer can then refreeze to form an overlying layer of *white ice* (Welch, Legault, and Bergmann 1987).

The formation of river ice is more varied, owing to the interactions between the growth of the ice and the dynamics of the river flow (Beltaos et al. 1993). The initial stage is again the formation of frazil, and along river banks this will normally produce congelation ice in the same way as for lake ice. In faster-flowing (hence likely to be turbulent) regions of the river, however, the frazil may be distributed throughout the water column, and may form masses of *anchor ice* adhering to the river bottom. The basic form of river ice can be considered to be a smooth transparent sheet of congelation ice. In the same way as for lake ice, this may have superimposed white ice, and it may also be interspersed with lumps of anchor ice that have been dislodged and have floated to the surface. Another important form of accumulation is *aufeis* or icing, typically where unfrozen river water is discharged onto the ice surface. Aufeis can also be formed from deep groundwater.

The basically smooth form of river ice can be disturbed by mechanical interaction with the river bed, by the presence of anchor ice or aufeis, and by mechanical interaction between ice floes that have been broken from the continuous cover, especially during the spring melt. Sequences of snowfall, melting and refreezing can give rise to a layered structure within the ice.

4.3.2   ELECTROMAGNETIC PROPERTIES OF FRESHWATER ICE IN THE OPTICAL AND NEAR-INFRARED REGIONS

The essential optical and near-infrared properties of black ice can be inferred from Figure 4.1; indeed, the reflection coefficient of below 2% throughout the optical region explains the origin of the name "black ice." Roughly speaking, we can say that such ice is transparent at wavelengths below about 1.5 μm and opaque at longer wavelengths. The reflectance properties of white ice are similar to those of snow, and hence represented by Figure 4.2. Absorption lengths in white ice are typically around 1 m at wavelengths below 0.6 μm, falling to around 1 cm at 1 μm (Perovich 1989).

4.3.3   THERMAL INFRARED EMISSIVITY OF FRESHWATER ICE

Figure 4.9 shows some recent measurements of the thermal infrared emissivity of freshwater ice. As one would expect, these are very similar to those for snow shown in Figure 4.4.

4.3.4   ELECTROMAGNETIC PROPERTIES OF FRESHWATER ICE IN THE MICROWAVE REGION

As discussed in Section 4.2.6., the real part of the dielectric constant of freshwater ice is 3.17 throughout the microwave region (Mätzler and Wegmüller 1987; Cumming 1952). The imaginary part depends on both frequency and temperature, and there is some variation in the reported values (Warren 1984; Ulaby, Moore, and Fung 1986; Hallikainen and Winebrenner 1992). However,

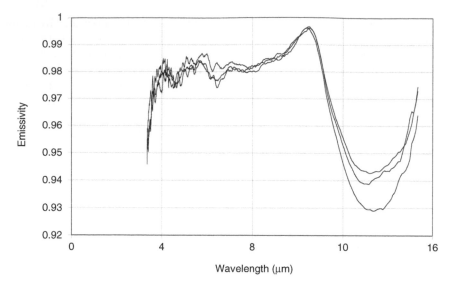

FIGURE 4.9 Three measurements of the spectral variation of the emissivity of freshwater ice between 3 and 15 μm (Zhang 1999).

TABLE 4.2
Predicted Values of the Imaginary Part of the Dielectric Constant of Freshwater Ice and the Corresponding Absorption Length, as a Function of Temperature and Frequency

Frequency (GHz)	−5°C		−15°C	
	ε''	Absorption Length (m)	ε''	Absorption Length (m)
1	0.0028	30	0.0014	61
3	0.0015	19	0.0008	36
10	0.0020	4.3	0.0013	6.5
30	0.0045	0.63	0.0036	0.79
100	0.0127	0.067	0.0120	0.071

the values are small, giving correspondingly long absorption lengths. As an example, Table 4.2 summarizes the predictions of the empirical model due to Hallikainen and Winebrenner (1992). A general empirical formula that takes both frequency f (in Hz) and temperature T (in K) into account is due to Nyfors (1982):

$$\varepsilon'' = 57.34\left(f^{-1} + 2.48 \times 10^{-14} f^{1/2} \right) \exp(0.0362T) \qquad (4.33)$$

Thus we would expect a typical cover of freshwater ice to be optically thin in the microwave region. Absorption is negligible, and the microwave scattering is dominated by reflections from the air–ice and ice–water interfaces. In the case of a very smooth-sided slab of ice, interference effects between the upper and lower surface reflections can occur.

This situation will be modified by roughness of the upper or lower surfaces of the ice, by the presence of inhomogeneities such as air bubbles within the ice, by white ice, a surface film of water, etc. Although clear ice has a dielectric constant of 3.17, milky ice (with a high concentration of bubbles) has been observed to have a dielectric constant of 3.08 and ice containing air bubbles larger than 6 mm to have a dielectric constant of 2.99 (Cooper, Mueller, and Schertler 1976).

Data on the passive microwave emissivity of freshwater ice are limited, probably because of the coarse spatial resolution of spaceborne passive microwave systems relative to the typical size of most bodies of river and lake ice (Eppler et al. 1992). At C-band (5 GHz), nadir brightness temperatures of smooth lake ice on Lake Erie are typically 195 to 210 K, increasing slightly with distance from land (Eppler et al. 1992; Swift, Harrington, and Thornton 1980). For rougher ice, including ridges, the brightness temperature generally exceeds 220 K. For comparison, open water has an emissivity of around 0.36 at this frequency, so that its brightness temperature would be approximately 100 K.

The microwave backscatter properties of freshwater ice are mainly dependent on its roughness, and hence on the manner of its formation. This is discussed in greater detail in Chapter 7.

4.4 SEA ICE

4.4.1 PHYSICAL CHARACTERIZATION

Sea ice is broadly similar to freshwater ice in its physical characterization, though some significant differences are introduced by the presence of salts and by the dynamic environment in which it exists. The phenomenology is very broad, and in this section we merely give a summary of the commonest and most important aspects. A more detailed review is given by, for example, Tucker et al. (1992).

The first major difference between sea ice and freshwater ice is that the freezing point of the former is lowered by the dissolved salts in the sea water, to around $-1.8°C$ for a typical salinity of 34 parts per thousand. For salinities above 25 parts per thousand, the temperature at which sea water attains its maximum density is actually less than the freezing point. As a consequence, the continued removal of heat from the water results in an unstable distribution of density, leading to convective overturning until the whole water column has reached the freezing point. The density and salinity structure of the ocean in fact limits this process to a surface layer a few tens of meters in depth.

As with freshwater ice, the first stage is the formation of frazil crystals. When the density of these crystals in the surface water is high enough to produce a viscous slurry it is referred to as *grease ice*; consolidation of this material results in a continuous sheet up to 10 cm thick, termed *nilas*, or less coherent *pancakes* of the order of a meter in diameter. Further thickening of the ice cover takes place mostly by congelation, as new ice is frozen onto the underside of the existing cover. The orientation of the growing crystals is controlled, as it is for freshwater ice (Weeks and Ackley 1986), and as the crystals form, brine is ejected and can be trapped between them in spaces known as *brine pockets*. The proportion of brine that is trapped in this way, rather than escaping from the growing ice sheet, depends on the growth rate in the sense that more rapidly growing ice covers will entrap more brine. Typical values in newly formed ice range from 5 to 12 parts per thousand (Maykut 1985), although as the ice ages it becomes progressively more desalinated as the brine escapes, largely by gravity drainage. Air bubbles are also entrapped in the ice as it forms.

First-year ice is sea ice that has not yet been exposed to the summer melting season. In the Antarctic, the thickness of first-year ice ranges from as little as 0.5 m to maybe 3 m in coastal bays. Ice less than 0.3 m thick is often referred to as *thin ice*, combining the categories of *new ice* and *young ice*. In the central Arctic Ocean, thicknesses of 1 to 2 m are common, although this can be increased by rafting (Drinkwater and Squire 1989) in the marginal ice zone. Ice that has survived at least one summer is termed *multiyear ice*, and it differs from first-year ice in a number of important respects. Desalination continues at an accelerated rate during the summer, especially in Arctic ice where the melting of the surface ice and any snow cover leads to flushing by fresh water. The bulk salinity of multiyear ice in the Arctic falls to around 3 parts per thousand; in the Antarctic the figure is typically twice as high. Other significant changes include the formation of *melt ponds* on the surface, which refreeze during the winter. A consequence of the formation of melt ponds is the development of a more irregular, *hummocky* surface.

Sea ice of both the first-year and multiyear varieties is kept in almost continuous motion by the action of the wind and of water currents. This can lead to the break-up of a continuous sheet into separate ice *floes*, the brittle or plastic deformation of floes as they are forced into collision with one another, producing coalescence with *ridges* (which can be of the order of 10 m high) along the boundary, and the further break-up of floes into *rubble*. Generally, the effect of these processes is to increase both the thickness and the irregularity of the ice cover as it ages, and to produce a scale-free distribution of floe sizes in which the number of floes $N(A)dA$ having areas between A and $A + dA$ is typically given by a power law:

$$N(A) = kA^{-n} \qquad (4.34)$$

where k is a constant that depends on weather conditions and n is typically 1.6 ± 0.2 (Korsnes et al. 2004). Linear areas of ice-free water, termed *leads*, can open up between floes. Two-dimensional areas of open water, termed *polynyas* (sometimes *polynyi*) can also develop (Martin et al. 1992).

The small-scale surface roughness of sea ice varies widely (Onstott 1992). For frazil, values of the RMS height variation σ range between 0.3 and 0.8 mm, with correlation lengths between about 5 and 15 mm. First-year congelation ice (the term is analogous to the definition for freshwater ice given in Section 4.3.1) is often not significantly rougher, with values of σ between 0.4 and 2 mm and correlation lengths between about 5 and 25 mm, although these values can range up to 10 mm and 80 mm, respectively, in the summer. Multiyear ice shows values of σ of between 1 and 10 mm, and correlation lengths of around 30 mm.

4.4.2 ELECTROMAGNETIC PROPERTIES OF SEA ICE IN THE OPTICAL AND NEAR-INFRARED REGIONS

The optical properties of young, growing, sea ice are markedly different from those of other types. Since the absorption properties of brine and of pure ice are more or less independent of temperature, the optical properties of young ice are governed by the effectiveness of scattering. The major influence is the included brine, followed in importance by air bubbles, growth rate, and temperature (Grenfell 1983). Figure 4.10 summarizes the spectral reflectance of a number of sea ice types. It illustrates the decline in reflectance from the blue end of the spectrum to the infrared, and the strong influence of air temperature. The reflectances remain low at longer infrared wavelengths, at least as far as 2.5 µm (Grenfell 1983). Absorption lengths at 0.4 µm range from about 1 m for fresh ice to 8 cm for multiyear ice, the primary determinant being the salinity. These values are more or less constant up to 0.6 µm but increase rapidly at longer wavelengths (Perovich and Grenfell 1981).

The data shown in Figure 4.10 all relate to ice about 20 cm thick. The reflectances are observed to increase with thickness.

4.4.3 THERMAL INFRARED EMISSIVITY OF SEA ICE

The thermal infrared properties of sea ice are similar to those of freshwater ice, i.e., the emissivity is typically 0.98.

4.4.4 ELECTROMAGNETIC PROPERTIES OF SEA ICE IN THE MICROWAVE REGION

The dielectric properties of sea ice in the microwave region are dependent on many factors, including the frequency, temperature, salinity, ice type, and fabric. A truly comprehensive investigation of the dependence on all of these factors

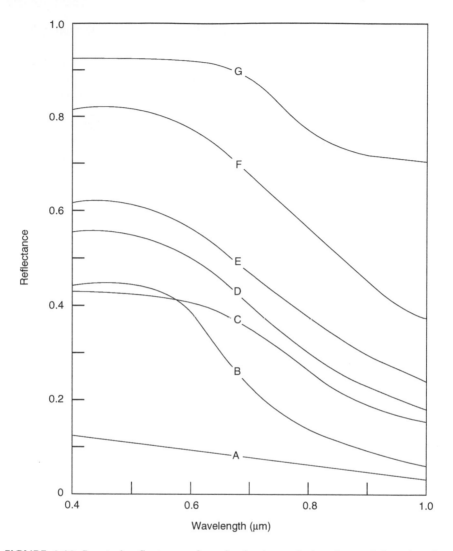

FIGURE 4.10 Spectral reflectance of sea ice in the optical and near-infrared region. A: fresh bubbly ice; B: melting first-year blue ice; C: full salinity first-year ice, $-10°C$; D: full salinity first-year ice, $-20°C$; E: full salinity first-year ice, $-30°C$; F: melting multiyear ice; G: full salinity first-year ice, $-37°C$. (Redrawn from Perovich and Grenfell 1981.)

would consequently require the collection of a very large amount of data, and has not yet been fully attempted. Nevertheless there is already a substantial body of knowledge, much of it reviewed by Hallikainen and Winebrenner (1992), and it is possible to make a few generalizations.

At microwave frequencies below about 15 GHz, the dominant variable is the salinity, followed by the temperature, with frequency playing a comparatively

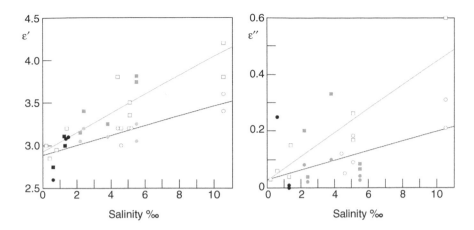

FIGURE 4.11 Real and imaginary parts of the dielectric constant of sea ice between 1 and 16 GHz. Black symbols are for −20°C, gray symbols for −5°C. Open circles represent first-year ice, solid circles multiyear ice, and squares artifically grown ice. The straight lines are linear regression fits to the data.

minor role. Figure 4.11 illustrates this point. It is based on data collected by Hallikainen and Winebrenner (1992) from many sources, and represents a range of ice types, including artificial, laboratory-grown ice, and frequencies between 0.9 and 16 GHz.

Figure 4.11 demonstrates that both the real and the imaginary parts of the dielectric constant increase with salinity. This is unsurprising, since the dielectric constant of seawater is very much larger than that of pure ice. For example, at a frequency of 10 GHz the dielectric constant of pure ice is almost entirely real and about 3.17, whereas that of seawater at 0°C is typically 42–47i. The figure also shows that the trends are more pronounced at −5°C than at −20°C, and in fact this dependence on temperature continues more rapidly as the temperature approaches the melting point.

The data in Figure 4.11 have been approximated by linear regression fits. The equations of these lines are

$$\begin{aligned}
\varepsilon' = 2.89 + 0.057s \quad &(T = -20°C) \\
2.93 + 0.112s \quad &(T = -5°C)
\end{aligned} \tag{4.35}$$

$$\begin{aligned}
\varepsilon'' = 0.030 + 0.017s \quad &(T = -20°C) \\
0.026 + 0.042s \quad &(T = -5°C)
\end{aligned} \tag{4.36}$$

although there is considerable scatter. In these equations, s is the salinity in parts per thousand. Typical values of the attenuation length at 37 GHz are of the order of 10 cm for multiyear ice and 1 cm for first-year ice. These values increase at lower frequencies.

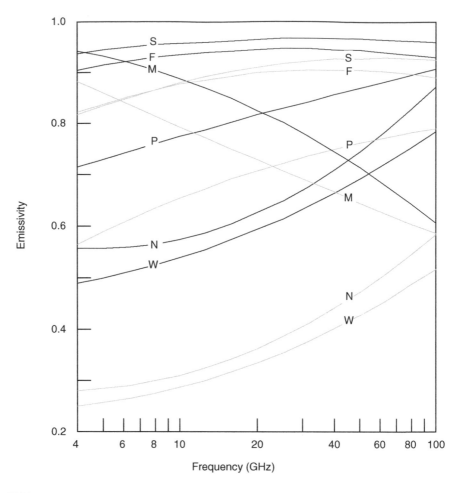

FIGURE 4.12 Emissivity of various types of sea ice at an incidence angle of 50°. The black curves are for vertical polarization and the gray curves for horizontal polarization. W: open water; N: new ice; P: pancake ice; M: multiyear ice; F: first-year ice; S: summer melting.

The emissivity of sea ice in the microwave region exhibits substantial variation with ice type. This is illustrated in Figure 4.12, which shows the emissivity spectra for both vertically and horizontally polarized radiation and for a range of ice types. The data have been plotted for an incidence angle of 50° rather than for normal viewing, since this is a common configuration for passive microwave radiometers. The curves in Figure 4.12 are based on data collected from many sources by Eppler et al. (1992), and have been derived by fitting second-order polynomials through the latter. They are subject to an uncertainty of typically ±0.04 in the emissivity.

Inspection of Figure 4.12 reveals a number of interesting facts. Firstly, we can note that the emissivity in vertically polarized radiation is greater than that in horizontally polarized radiation at all frequencies in the range 4 to 100 GHz for all ice types. This is expected for materials in which the emission is predominantly from the surface, since the Fresnel reflection coefficient of such materials is greater for horizontally polarized than for vertically polarized radiation. We also note that, with the exception of first-year and melting ice, the emissivities show strong trends with frequency — increasing, except in the case of multiyear ice. These differences in behavior point to the possibility of using multifrequency (and dual-polarization) measurements to discriminate among different ice types.

The microwave backscattering properties of sea ice are complicated and dependent on a large number of variables, including ice type and structure, presence and condition of a snow cover, and environmental conditions, in addition to the imaging parameters (frequency, polarization state, and incidence angle) of the radar. Onstott (1992) has presented a thorough review, and further data are given by Lewis et al. (1994). Figure 4.13, which is derived from data presented by Onstott (1992), summarizes in schematic form the typical variation of backscattering coefficient with incidence angle for first-year and multiyear ice. It also shows the backscattering behavior of calm open water. The data in Figure 4.13 are for HH-polarized observations: the results for VV-polarized observations are broadly similar.

4.5 GLACIERS

4.5.1 PHYSICAL CHARACTERIZATION

A glacier is a large (usually larger than 10 hectares) mass of ice having its origin on land, and normally displaying some movement. It is usually in a state of approximate dynamic equilibrium, with a net input of material in its upper *accumulation area* and a net loss of material in its lower *ablation area*. These two areas are separated by the *equilibrium line*. Figure 4.14 illustrates schematically the relationship between these areas and the flow of ice within the glacier. The input to a glacier is in the form of snowfall, and the output is principally in the form of meltwater and, in the case of a glacier whose lower terminus is in water, icebergs that calve off and float away. The snow is transformed to ice through a number of mechanisms.

The surface of a glacier can be divided into a number of zones or facies, an idea developed by Benson (1961) and Müller (1962). These are illustrated in Figure 4.15, which is adapted from Paterson (1994), as is the following description. Uppermost is the *dry-snow zone* in which no melting takes place. This zone is only found inland in Greenland* and over most of the Antarctic

---

*About a quarter of the surface of the Greenland ice sheet experiences summer melting (Abdalati and Steffen 2001).

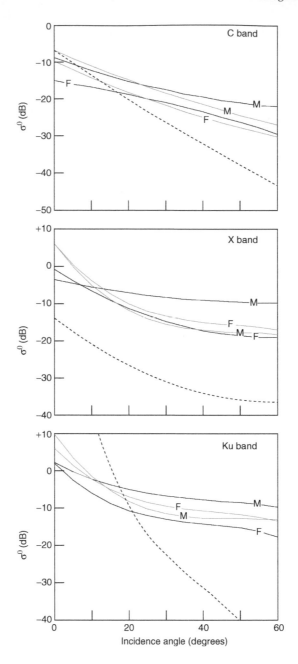

FIGURE 4.13 Typical copolarized (HH) backscatter coefficients of first-year (F) and multiyear (M) sea ice as a function of incidence angle. The black curves are for winter and the gray curves for summer. The dotted lines represent calm open water. The panels show three different frequencies: C band (5 GHz), X band (10 GHz), and Ku band (13 GHz).

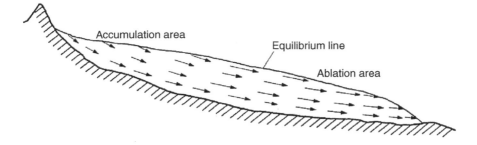

FIGURE 4.14 Schematic diagram of an idealized glacier. (Reprinted from Paterson (1994). Copyright 1994, with permission of Elsevier.)

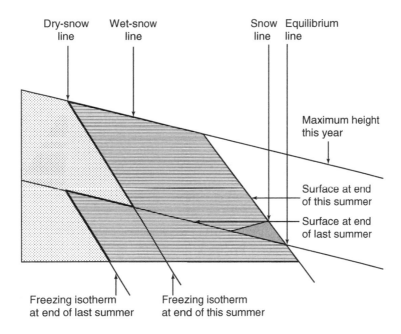

FIGURE 4.15 Schematic cross-section through a glacier, illustrating how its surface is separated by the dry snow, wet snow, snow, and equilibrium lines. Stippling denotes snow, horizontal shading firn, and vertical shading superimposed ice.

ice sheet, and on the highest mountain glaciers, where the annual average temperature is lower than some threshold value that was originally proposed as −25°C (Benson 1961) but is now believed to be around −11°C (Peel 1992). Next is the *percolation zone*, in which some surface melting occurs during the summer. The meltwater percolates downward and refreezes to form inclusions of ice in the form of layers, lenses, and pipes. The dry-snow and percolation zones are separated by the *dry-snow line*. Below the percolation zone, and separated from it by the *wet-snow line*, is the *wet-snow zone* in which all the current year's snowfall melts. Below this, and separated from it by the *snow line* (sometimes called the *firn line*) is the *superimposed ice zone*. In this zone, surface melting is so extensive that the meltwater refreezes into a continuous mass of ice. The lower boundary of this zone is the equilibrium line.

Not all of these zones are present in all glaciers. As has already been stated, only the coldest glaciers possess dry-snow zones. Temperate glaciers, in which the temperature of all but the upper few meters is at the freezing point, exhibit only wet-snow and ablation zones (the superimposed ice zone is generally negligible so the snow line and equilibrium line coincide). *Ice shelves*, which are parts of glaciers or ice sheets that extend over water, do not have ablation zones but instead lose material mostly by calving of glaciers.

Snow is transformed to ice in a glacier by a variety of mechanisms. These include mechanical settling, sintering (in the dry-snow zone), refreezing of meltwater, and refreezing of sublimated ice to form *depth hoar*. The density of the material in a glacier increases with depth. Once the transformation process has begun, the material is referred to as *firn* rather than as snow, and firn (also called *névé*) generally has a density greater than $0.55\,\mathrm{Mg\,m^{-3}}$. Firn is porous, since it contains interconnected air channels. However, once the density increases above about $0.83\,\mathrm{Mg\,m^{-3}}$ these channels are closed off, resulting in ice in which closed air bubbles are trapped. Grain sizes in glaciers generally increase with depth, from typically 0.5 to 1 mm near the surface to a few millimeters at greater depths. The grains in depth hoar can be up to 5 mm in size.

4.5.2   ELECTROMAGNETIC PROPERTIES OF GLACIERS IN THE OPTICAL AND NEAR-INFRARED REGIONS

In the winter, a glacier surface is usually covered by snow. The optical properties of snow cover have already been discussed in Section 4.2.4. In the summer, however, other surfaces can be exposed. Figure 4.16 summarizes some experimental data on the spectral reflectance properties of glacier surfaces. The spectra *a*, *c*, *e*, and *f*, for fresh snow, firn, clean glacier ice, and dirty glacier ice, respectively, are adapted from Qunzhu, Meishing, and Xuezhi (1984), while the spectra *b* and *d*, for the accumulation and ablation areas of Forbindels Glacier in Greenland, are adapted from Hall et al. (1990). Comparison of curve *a* with Figure 4.2 shows that the snow grain size in this case is about 0.2 mm.

We note from Figure 4.16 a general tendency for the visible-wavelength reflectance of a glacier surface in summer to increase with altitude, moving upwards from the ablation area. This is illustrated in Figure 4.17, which shows the broad-band albedo of the glacier Midre Lovénbreen in Svalbard, measured *in situ* in summer. At wavelengths longer than about 600 nm the same general trend is observed, although there is greater observed scatter in the results,

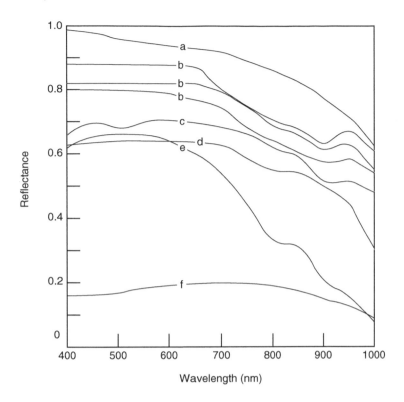

FIGURE 4.16 Spectral reflectance of different glacier surfaces (simplified). a: fresh snow; b: accumulation area; c: firn; d, e: glacier ice; f: dirty glacier ice. See the text for further details.

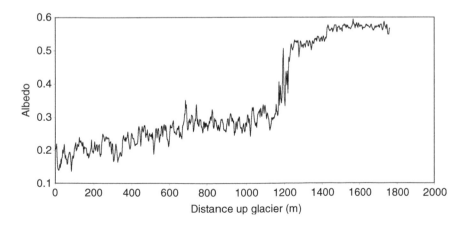

FIGURE 4.17 Longitudinal profile of Midre Lovénbreen, Svalbard, showing the variation of summer albedo. (Data supplied by Dr. N. S. Arnold, Scott Polar Research Institute.)

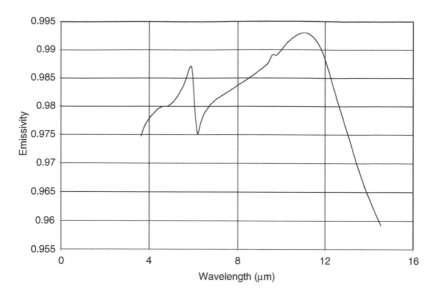

FIGURE 4.18 Emissivity spectrum of pure (distilled) water (Zhang 1999).

presumably as a consequence of the greater sensitivity to structural details such as grain size (Winther 1993) and the quantity and size of air bubbles in ice.

4.5.3 THERMAL INFRARED EMISSIVITY OF GLACIERS

Figures 4.4 and 4.9 show the thermal infrared emissivity of snow and ice, respectively. For comparison, Figure 4.18 shows the emissivity spectrum of pure water. We note that all emissivities are very similar (0.993 to 0.998) at around 11 μm, although significant differences occur away from this wavelength.

4.5.4 MICROWAVE PROPERTIES OF GLACIERS

The microwave backscattering properties of most glaciers show significant seasonal variations (Rott, Nagler, and Floricioiu 1995; Rees, Dowdeswell, and Diament 1995). During the summer, the accumulation area (excepting the dry-snow area if present) is generally covered with wet snow, and scattering from this area is dominated by surface scattering (Marshall, Rees, and Dowdeswell 1995). It is therefore controlled by the water content of the snow, which determines the dielectric constant, and the surface roughness, as discussed in Section 4.2.6. Usually the snow surface is smooth, so the backscattering coefficient is low unless the incidence angle is also low (i.e., unless the radar is viewing almost perpendicularly to the surface), and there is a significant correlation between backscatter and surface topography (Dowdeswell, Rees, and Diament 1993). Surface scattering also dominates in the ablation area, and is controlled primarily by the surface roughness. This is generally higher than over the snow-covered areas (Brown, Kirkbride, and Vaughan 1999), and increases throughout the summer as a result of melting. Crevassed areas can give high values of backscatter, especially if the crevasses are oriented perpendicular to the look-direction of the radar (Rees, Dowdeswell, and Diament 1995).

During the winter, the snow cover is dry and hence practically transparent to microwave radiation (Section 4.2.6). In the ablation area the backscattering is dominated by scattering from the interface between the ice and the overlying dry snow. In the accumulation area, volume scattering dominates, the effective volume scatterers being the ice grains but especially any ice layers, lenses, or pipes that are present (Rott, Sturm, and Miller 1992; Rott, Nagler, and Floricioiu 1995; Engeset and Ødegård 1999). Backscattering coefficients are high except in the dry-snow zone (which does not contain any large inclusions of ice). The transition from summer to winter conditions is thus characterized by a large increase (typically 10 dB or more) in backscatter over the firn, although this occurs a few weeks later than freezing of the surface as a result of the time delay in conducting heat away from the deeper layers of the glacier (Rott, Nagler, and Floricioiu 1995). Backscatter coefficients over dry-snow areas are generally low (typically −10 dB at C band) compared with the percolation and wet-snow zones. Bindschadler, Fahnestock, and Kwok (1992) and Forster et al. (1999) have demonstrated a significant relationship between the backscatter coefficient and the accumulation rate, in the sense that higher accumulation rates are associated with lower values of the backscatter coefficient.

Figure 4.19 summarizes data on the backscattering coefficient of glaciers at C band (ca 5 GHz) and L band (ca 1.3 GHz), for both copolarized and cross-polarized observations. The full curves are smoothed representations of data presented by Rott, Nagler, and Floricioiu (1995) from the Ötztal glacier in Austria. The single points are averages of data obtained by Rees, Dowdeswell, and Diament (1995) from Nordaustlandet, Svalbard and by Engeset and Ødegård (1999) from Slakbreen on Svalbard. It should be borne in mind that these data are merely indicative, since surface roughness and the distribution of ice layers, etc. will vary significantly from one glacier to another.

The microwave emissivity characteristics of glaciers have been largely discussed in Section 4.2.7 (microwave emissivity of snow).

4.5.5 TRANSMISSION OF VHF AND UHF RADIATION THROUGH GLACIERS

The VHF (very high frequency: 30 to 300 MHz) and UHF (ultrahigh frequency: 300 MHz to 3 GHz) bands are not normally of much importance

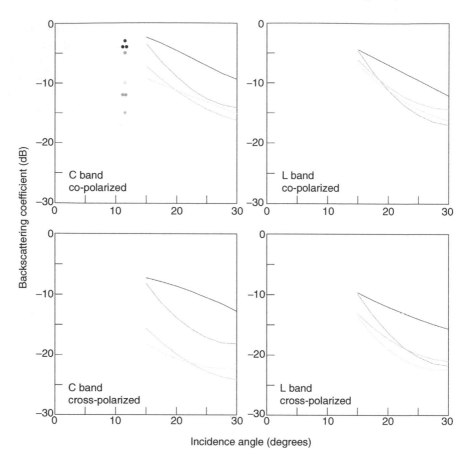

FIGURE 4.19 Backscattering coefficient of glaciers as a function of incidence angle. Black: winter; gray: summer; full curves: accumulation area; dotted curves: ablation area.

in remote sensing, since the long wavelengths impose low angular resolution. They do, however, have a major role to play in the study of glaciers, since in this region of the electromagnetic spectrum, ice and snow are practically transparent. This forms the basis of the technique known as *radio echosounding* for measuring the depth of glaciers (Macqueen 1988; Bogorodsky, Bentley, and Gudmandsen 1985), and the similar technique of *groundpenetrating radar* (GPR) when applied to glaciers. This is a pulsed technique, in which short pulses of electromagnetic radiation are transmitted downward into the glacier. The time delay of the return signal gives the depth of the glacier, and weaker intermediate returns can give information on its stratification. Conversion between the time delay and the depth requires that the propagation speed, or equivalently the refractive index, be known. This depends strongly on density, weakly on temperature, and negligibly on

TABLE 4.3
Iceberg Classification According to the International Ice Patrol (IIP)

Type	Freeboard (m)	Width (m)	Typical Mass (t)
Growler	< 1.5	< 5	10^3
Bergy bit	1.5–5	5–15	10^4
Small berg	5–15	15–60	10^5
Medium berg	15–50	60–120	10^6
Large berg	50–100	120–220	10^7
Very large berg	> 100	> 220	> 10^7

From Willis et al. (1996).

frequency throughout the VHF and UHF bands. The density-dependence of the refractive index n on the density ρ can be described by

$$n = 1 + K\rho \qquad (4.37)$$

where the constant K is around $(8.4 \pm 0.1) \times 10^{-4} \, \text{m}^3 \, \text{kg}^{-1}$ (Rees and Donovan 1992). A value of $8.45 \times 10^{-4} \, \text{m}^3 \, \text{kg}^{-1}$ has been proposed by Kovacs, Gow, and Morey (1995). Discussion of the optimal technique for determining the propagation speed and density is presented by Eisen et al. (2002).

4.6   ICEBERGS

4.6.1   Physical Characterization

Icebergs are broken-off ("calved") pieces from the end of a glacier or ice shelf that terminates in water, or which have calved from other icebergs. They are classified by both size and shape, and two systems of classification are in widespread use (Willis et al. 1996). The World Meteorological Organization (WMO) system is based principally on shape, and defines *icebergs*, *bergy bits*, and *growlers*. Icebergs, with freeboards* greater than 5 m, are subdivided into tabular, dome-shaped, pinnacled, weathered, and glacier bergs. Bergy bits, with freeboards between 1 and 5 m, usually have areas between 100 and 300 m², while growlers have freeboards of less than 1 m and areas of typically 20 m². The International Ice Patrol (IIP) classification system is based purely on size, and is summarized in Table 4.3.

The very largest *tabular* icebergs, sometimes referred to as *ice islands* (although this term is sometimes used only to refer to cases in which the

---

*The freeboard is the height above the water line.

icebergs are grounded), calve from the Antarctic ice shelves, and can be tens or even hundreds of kilometers in width. Tabular icebergs also occur in the Arctic, particularly through calving from the ice shelves of Ellesmere Island (Koenig et al. 1952).

The morphology and form of the source glacier influence the size and shape of the resulting icebergs (Dowdeswell 1989; Løset and Carstens 1993) and indeed the internal structure, and further changes occur during the iceberg's life as it melts and undergoes fragmentation and erosion. If a tabular iceberg does not capsize, it will initially preserve the characteristics of the ice shelf from which it calved, with a relatively flat (though possibly heavily crevassed) surface and steep sides. Smaller icebergs, especially bergy bits and growlers, tend to be more irregular in shape.

4.6.2 ELECTROMAGNETIC PROPERTIES OF ICEBERGS IN THE OPTICAL AND INFRARED REGIONS

Considered as large pieces of glacier ice or firn, the optical and infrared properties of icebergs have already been described. The thickness of an iceberg greatly exceeds the attenuation length of the radiation, so the iceberg is "optically thick" (opaque).

4.6.3 ELECTROMAGNETIC PROPERTIES OF ICEBERGS IN THE MICROWAVE REGION

Similarly, the essential principles of microwave interactions with icebergs have already been described in considering the microwave properties of glaciers. Microwave backscattering from the surface of an iceberg is comparatively weak, since the dielectric constant of freshwater ice is fairly small (3.15 to 3.20). Most of the incident energy in a radar observation will thus enter the bulk of the iceberg, to undergo volume scattering from the many air bubbles trapped within the ice, as well as from other inhomogeneities including internal cracks. At sufficiently low frequencies, or for sufficiently small icebergs, radar radiation may pass through the entire iceberg and undergo surface scattering from its bottom (Gray and Arsenault 1991). Another important mechanism in the interaction of radar radiation with icebergs is the possibility of a "double bounce" process, in which radiation is first scattered from the wall of the iceberg and then from the surface of the water in which it is floating (or these bounces can occur in the reverse order). This can produce a strong specular scattering, although it is sensitive to the orientation of the iceberg, since it depends on the iceberg presenting a wall that is (1) perpendicular to the water surface and (2) facing the radar. As a multiple-scattering process, it contributes a significant cross-polarized element to the backscattering cross-section of the iceberg (Kirby and Lowry 1979).

5 Remote Sensing of Snow Cover

5.1 INTRODUCTION

As was observed in Section 1.2, snow cover is a large-scale phenomenon. The maximum winter extent in the northern hemisphere is about 46 million square kilometers, but the spatial density of the *in situ* measurement network is too low to provide an adequate characterization of its distribution (Mognard 2003). Remote sensing of snow cover is therefore desirable, and is in fact well developed. The most important types of imagery for this application are optical (including the use of near-infrared) and passive microwave imagery (although synthetic aperture radar is also being developed as a method), since in both of these regions of the electromagnetic spectrum snow has distinctive properties that make it comparatively easy to detect. In the visible–near-infrared region, snow has a very high albedo (see Color Figures 2.4 and 5.1 following page 108). In the microwave region, snow has a brightness temperature that is significantly lower than that of snow-free ground, because of volume scattering in the snowpack. Snow is also practically unique among land cover materials in that its emissivity decreases with frequency (Mätzler 1994).

Snow was observed in the first image obtained from the TIROS-1 satellite in 1960 (Singer and Popham 1963). Snow cover has been routinely monitored from space using optical imagery since 1966 (Matson, Ropelewski, and Varnardore 1986), and since 1978 using passive microwave imagery (Hall et al. 2002). The two approaches have somewhat complementary advantages and disadvantages. While the optical systems (Frei and Robinson 1999) can achieve much higher spatial resolutions, they are limited to daylight and cloud-free conditions. This is a particular difficulty in regions with maritime climates, such as the United Kingdom, where cloud cover is frequently associated with snow cover (Archer et al. 1994). Passive microwave systems can be used at night and through cloud, but their spatial resolution is coarse. These resolutions are generally too coarse for hydrological applications (Standley 1997) where optical and SAR data may be preferable.

A problem common to all approaches to remote sensing of snow cover is that much of the world's seasonal snow cover occurs in forested areas (Hall et al. 2001) and complex landscapes generally (Walker and Goodison 1993; Solberg et al. 1997; Vikhamar and Solberg 2000; Klein, Hall, and Riggs 1998; Stähli, Schaper, and Papritz 2002).

FIGURE 5.1 Extract of visible wavelengths of a Landsat 7 ETM+ image showing Minneapolis and St Paul, Minnesota, on 15 March 2001. The heavy snow cover caused extensive flooding. (Image courtesy of EROS Data Center and Landsat 7 science team, downloaded from NASA Visible Earth web site at http://visibleearth.nasa.gov.)

5.2   SPATIAL EXTENT

The following discussion of monitoring the extent of snow cover is organized according to spatial scale.

5.2.1   SMALL SCALE

5.2.1.1   VIR Imagery

At the finest spatial scales, typical of the spatial resolution of sensors such as Landsat (of the order of a few tens of meters), most approaches to the monitoring of snow cover are based on the analysis of VIR imagery, although SAR also provides a possible technique. As pointed out in Sections 4.2.4 and 5.1, snow has a very high reflectance, at least for wavelengths below about 0.8 μm, so it is comparatively easy to detect against a background of snow-free terrain (Figure 5.1). At the most basic level, single-band imagery can simply be thresholded to discriminate between snow-free and snow-covered terrain. However, the choice of a suitable threshold value is not always straightforward, and ancillary information may be needed. The threshold may depend on the local imaging geometry, since surfaces that face squarely toward the incident solar radiation will reflect more light to the sensor than those that are oriented more obliquely. A further disadvantage of using single-band imagery is the potential for confusion introduced by cloud cover.

A commoner approach is to use multispectral imagery. Most usefully, this consists of a spectral band in the visible part of the spectrum, plus one centered near 1.65 μm in the near infrared. The reason for this is that, while the spectral reflectances of snow cover and of cloud are very similar at wavelengths below about 1 μm (Massom 1991; König, Winther, and Isaksson 2001), they diverge in the near infrared and achieve a maximum difference (in the sense that cloud is more reflective than snow) at wavelengths between about 1.55 and 1.75 μm* (Figure 5.2). Furthermore, the use of two spectral bands allows for a correction of the topographic effect referred to in the previous paragraph. Variations in the relative sun–ground surface–sensor geometry will (to first order) change the at-sensor radiance by the same factor in all spectral bands, so that the ratio of the measured radiance in two spectral bands will be unaffected by the geometry. The ratio is not in fact the most convenient quantity to define, since it can, in principle, take any value between zero and infinity. Instead, a *normalized difference index* is defined. For the discrimination of snow, the usual index is called the *normalized difference snow index* (NDSI),[†] defined in terms of the spectral bands of Landsat TM and ETM+ as

$$\text{NDSI} = \frac{r_2 - r_5}{r_2 + r_5} \qquad (5.1)$$

where r_2 and r_5 are the reflectances in bands 2 and 5 (center wavelengths 0.57 μm and 1.65 μm), respectively (Dozier 1984, 1989). Snow is normally assumed to be present if the NDSI exceeds a value of 0.4 (Dozier 1989; Hall, Riggs, and Salomonson 1995), although recent work has suggested that the optimum value of the threshold varies seasonally. Based on field investigations around Abisko, Sweden, Vogel (2002) has suggested that a threshold of 0.48 is more appropriate in July, 0.6 in September.

A combination of multispectral analysis and simple thresholding gives the possibility of extending the analysis of snow cover to higher spatial resolutions, as demonstrated by Vogel (2002). As was described earlier, the principal

---

*Thin cirrus cloud, however, remains difficult to discriminate from snow cover in this way (Hall, Riggs, and Salomonson 1995).

[†]The NDSI is a particular example of the general class of normalized difference indices, of which the NDVI (normalized difference vegetation index) is the original and most famous. It is defined similarly to the NDSI but with a near-infrared reflectance (e.g., from TM or ETM+ band 4) taking the place of the band 2 reflectance and a red reflectance (e.g., TM or ETM+ band 3) taking the place of band 5.

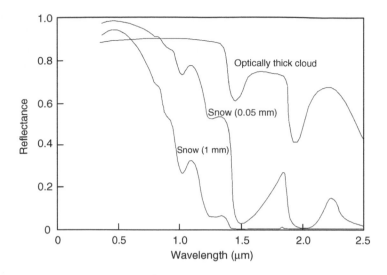

FIGURE 5.2 Spectral reflectance of optically thick cloud compared with snow cover of two different grain sizes.

difficulty in applying the thresholding technique to single-band data is that the threshold may vary spatially and indeed temporally. However, by using suitable multispectral imagery to identify snow-covered areas against a background of snow-free terrain, a local threshold can be defined for higher-resolution single-band imagery. This is an attractive proposition for the ETM+ imagery recorded by the Landsat 7 satellite, since it provides multispectral imagery at a resolution of 30 m and a single panchromatic band (band 8) with a resolution of 15 m. By combining an analysis of the NDSI (from bands 2 and 5) with visual inspection of a multispectral (bands 2, 4, and 5) composite, Vogel (2002) was able to set a local threshold for the band 8 data and hence map smaller snowfields at a resolution of 15 m. The presence of shadows introduces a difficulty, although this can be resolved using a digital elevation model (DEM) of the terrain to model and allow for the occurrence of shadows (Baral and Gupta 1997; Datcu 1997).

5.2.1.2   SAR Imagery

The potential of synthetic aperture radar imagery for monitoring snow cover was discussed as early as 1980 (Goodison, Waterman, and Langham 1980), recognizing its advantages of high spatial resolution, insensitivity to cloud, and operability at night, but the relevant electromagnetic interactions are complicated. The attenuation length of microwave radiation in cold dry snow is large (Section 4.2.6), and this kind of snow is therefore transparent and hence invisible to radar (Rott and Sturm 1991) unless the snow pack is very deep (or at radar frequencies above about 10 GHz, which are not commonly used for SAR imaging). However, when the liquid water content of snow exceeds about 1% the attenuation length is reduced to a few centimeters, and the radar backscatter is usually dominated by surface scattering (Ulaby, Stiles, and Abdelrazik 1984; Rott and Sturm 1991). However, the question of whether such snow can be discriminated from snow-free terrain depends on the geometric and electromagnetic characteristics of the snow cover and the snow-free terrain. Various studies have demonstrated that wet-snow cover can generally be distinguished from snow-free terrain (Rott 1984; Hallikainen et al. 1992; Rott and Nagler 1995; Baghdadi, Fortin, and Bernier 1999; Rees and Steel 2001). The discriminatory ability increases with frequency so that observations at X-band are preferred, although C-band imagery is usually adequate (Shi and Dozier 1993). In most cases, the C-band copolarized backscatter at an incidence angle of about 23° (i.e., typical of the ERS instruments) is typically 2 to 3 dB lower for wet snow than for snow-free terrain (Hallikainen et al. 1992; Koskinen et al. 1994; Rees and Steel 2001) (Figure 5.3). However, there is no guarantee that this will be so and in some cases wet snow and snow-free terrain have proved to be indistinguishable (Haefner et al. 1993; Shi and Dozier 1997). This is true for densely forested areas for example (Koskinen, Pulliainen, and Hallikainen 1997; Baghdadi, Gauthier, and Bernier 1997; Rees and Steel 2001). Approaches to modeling the contribution from forested areas have been proposed by Koskinen, Pulliainen, and Hallikainen (1997). There is some evidence that discrimination increases at higher incidence angles (Guneriussen 1998; Baghdadi, Livingstone, and Bernier 1998).

The use of SAR imagery for snow cover detection generally proceeds in one of two ways. Multitemporal imagery can be used to establish the reference, snow-free, backscatter as a function of location (Haefner and Piesbergen 1997;

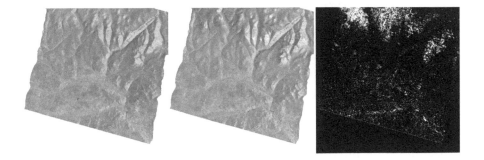

FIGURE 5.3 Detectability of wet-snow cover in ERS-2 (CHH) SAR imagery. The images show a 28.5 km square area of the Scottish highlands, and have been corrected for layover. The left image is a reference image (3 February 1997) showing snow-free conditions. The center image (15 March 1999) was acquired during a period of wet-snow cover. The white areas in the right image show areas where the backscattering coefficient has decreased by at least 2 dB, corresponding to areas of snow cover. (ERS-2 images. Copyright ESA 1997, 1999. Data from author's research.)

Rott and Nagler 1993). The optimum situation for this reference image is a winter image with frozen soil overlain by dry snow, so that undesirable variations in backscatter are not introduced by variations in soil moisture (Baghdadi, Fortin, and Bernier 1999). This reference image can then be used to set a spatially dependent threshold for the presence of wet snow. The algorithm proposed by Baghdadi, Fortin, and Bernier (1999) is to infer that wet snow is present if

$$\sigma_r^0 - \sigma^0 \geq a \text{ AND } b \leq \sigma^0 \leq c \tag{5.2}$$

where σ^0 is the observed backscatter coefficient, σ_r^0 is the reference (snow-free or dry-snow) backscatter coefficient, and a, b, and c are model parameters based on experimental observations. Alternatively, an image can be corrected for variations in incidence angle using an empirical model of the dependence of backscatter on incidence angle, provided that a sufficiently accurate and high-resolution digital elevation model is available. The corrected image is then thresholded to identify areas of wet-snow cover. In areas where the underlying land cover is heterogeneous, it is probably necessary to establish different thresholds for different land cover types, which requires that a land cover map is also available (Rees and Steel 2001). This is a simple example of a data synergy, where the land cover map can be produced, for example, from multispectral classification of a VIR image.

Most studies to date have demonstrated the limitations imposed by mono-spectral SAR at a single incidence angle. Multipolarization and/or multi-frequency SAR shows promise of greater discrimination (Shi and Dozier 1997; Narayanan and Jackson 1994), as does the use of InSAR techniques (Shi and Dozier 1997; Strozzi, Wegmuller, and Mätzler 1999) (wet snow exhibits metamorphism that reduces the coherence between images). In particular, polarimetric SAR allows snow extent and physical characteristics to be derived in a manner that is much less sensitive to topographic variations than single-polarization SAR (Shi, Dozier, and Rott 1994).

5.2.2 Intermediate Scale

At intermediate scales (spatial resolutions of the order of 1 km), the principles of mapping snow extent from remotely sensed data are similar to those at fine scales. Increased difficulties are caused by the fact that pixels representing larger areas of the Earth's surface are more likely to contain a mixture of land cover types. On the other hand, relaxing the requirement for high spatial resolution increases the range of sensors that have potential to supply useful data. Furthermore, coarser-resolution imagery is usually acquired from a wider swath, which increases the potential temporal resolution of the data. It is thus usually easier to capture transient or rapidly varying phenomena in coarser-resolution imagery. Even so, the presence of cloud cover often prevents the acquisition of data, especially in areas of temporary snow cover (Archer et al. 1994).

The archetypal satellite imaging system for intermediate-scale mapping of snow cover is the AVHRR (Advanced Very High Resolution Radiometer). This system has been in operation since 1979, and the more recent generations of the instrument, since 1998, have included a band centered at 1.6 μm, which is valuable for discrimination of snow from cloud. Regional snow monitoring programs using AVHRR imagery have been widely implemented, for example, through the ASCAS (Alpine Snow Cover Analysis System) in the Alps (Baumgartner and Apfl 1993; Baumgartner, Apfl, and Holzer 1994) and also in Greenland (Hansen and Mosbech 1994), Norway (Solberg and Andersen 1994) and the United Kingdom (Harrison and Lucas 1989; Xu et al. 1993). The broadly similar MSU-S instrument has been employed, together with AVHRR, in Russia (Burakov et al., 1996).

Two main approaches have been adopted for mapping snow cover from AVHRR data: unsupervised image classification followed by manual classification against a false-color composite, and linear interpolation based on NDVI values. In the latter approach (Slater et al. 1999), the fractional snow cover of a pixel is estimated as

$$\frac{N_c - N_w}{N_s - N_w} \tag{5.3}$$

where N_c is the current value of NDVI, N_w is the maximum value for snow-free vegetation (which may require seasonal adjustment), and N_s is the value for snow cover (usually taken to be zero).

A broadly similar instrument to the AVHRR is VEGETATION,[*] carried on the SPOT series of satellites since 1998. Like the AVHRR, this instrument lacks spectral coverage equivalent to Landsat band 2. An alternative Normalized Difference Snow and Ice Index (NDSII) has been proposed by Xiao, Shen, and Qin (2001). In terms of Landsat TM bands, this is defined as

$$\text{NDSII} = \frac{r_3 - r_5}{r_3 + r_5} \tag{5.4}$$

so that it employs the ratio of the band 3 (red: 0.63 to 0.69 μm) to band 5 (mid infrared: 1.55 to 1.75 μm) reflectances. It can also be adapted to the VEGETATION instrument.

Since the year 2000, data have been available from the MODIS sensor[†] carried on the Terra satellite (Figure 5.4). This represents an improvement on AVHRR from the point of view of snow mapping, since it has a finer spatial resolution (250 or 500 m compared with 1.1 km) and more spectral bands.

---

[*]http://www.spot-vegetation.com/.
[†]http://modis-snow-ice.gsfc.nasa.gov/snow.html.

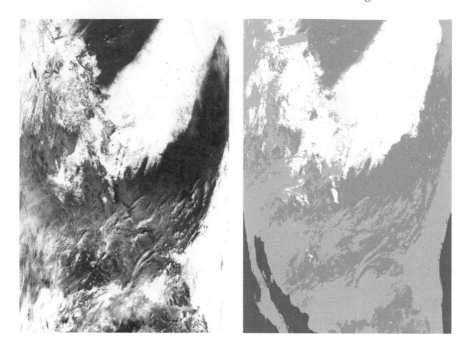

FIGURE 5.4 (See Color Figure 5.2 following p. 108) MODIS image of the midwestern United States, 1 February 2002 (http://modis-snow-ice.gsfc.nasa.gov/MOD10_L2.html). (Left) multispectral composite; (right) corresponding classified snow cover image, showing snow, cloud, snow-free land, and water, as progressively darker shades of gray.

Snow mapping from MODIS data uses bands 2, 4, and 6 (0.86, 0.56, and 1.64 μm, respectively) to determine whether snow is present or not, and additionally uses a vegetation index calculated from bands 1 (0.64 μm) and 2 to improve detectability in dense forests (Klein, Hall, and Riggs 1998). The MODIS version of the NDSI is defined as

$$\text{NDSI} = \frac{r_4 - r_6}{r_4 + r_6} \tag{5.5}$$

where r_4 denotes the reflectance in band 4, and similarly, so it is virtually identical to the Landsat NDSI defined by equation (5.1). The criterion for snow cover in an area not densely forested is

$$\text{NDSI} \geq 0.4 \text{ AND } r_2 \geq 0.11 \text{ AND } r_4 \geq 0.10 \tag{5.6}$$

(Hall et al. 2002), where the thresholds in bands 2 and 4 help to eliminate areas of water, which can also show high values of NDSI. This algorithm generally performs well, though it seems to overestimate snow cover in areas of patchy snow. This may be a consequence of the "binning" scheme.

As was mentioned earlier, snow cover often occurs in areas of complex landscape, for example, including areas that are forested as well as clear of trees. At coarser spatial resolutions, the problem of "mixed pixels" increases. One approach to this problem of complex landscapes is the use of spectral mixture modeling ("spectral unmixing") of imagery with a sufficiently high spectral resolution. This has been applied successfully by Nolin, Dozier, and Mertes (1993) using airborne AVIRIS data. This instrument has 224 spectral channels between 400 and 2460 nm. The technique consists of identifying pixels in the image representing "end-members" that are believed to represent homogeneous land cover types. The reflectance spectra of these end-members are determined. Other pixels in the image are assumed to consist of linear mixtures of these pure land cover types, so that the reflectance r_i observed in spectral band i is modeled as

$$r_i = \sum_{j=1}^{N} f_j r_{ij} + e_i \tag{5.7}$$

where f_j is the fraction of land cover type j present in the pixel, r_{ij} is the band-i reflectance of land cover type j, and e_i is the error in the linear model. N is the total number of land cover classes. The spectral unmixing process involves determining the set of coefficients f_j that minimize the sum of the squares of the errors e_i. A similar approach using Landsat TM imagery has been described by Rosenthal and Dozier (1996).

5.2.3   GLOBAL SCALE

Mapping and monitoring of snow cover on a national or global scale can be achieved using the "intermediate scale" techniques discussed in the previous section. However, the further relaxation of the requirement to achieve high spatial resolution that is implicit in the larger area of coverage introduces the possibility of using passive microwave remote sensing. As has already been noted, the spatial resolution achievable from spaceborne passive microwave observations is coarse (e.g., the footprint of the 19 GHz band of the SSM/I is 70×45 km), and this imposes a significant problem for one of the major potential uses of global snow cover data, in hydrology. Use of the SSM/I band at 85 GHz has attracted some interest for this reason. Although the footprint size is nominally about 15 km, with a sampling interval of 12.5 km, oversampling and deconvolution allow a resolution of about 5 km to be extracted from the data (Standley and Barrett 1994, 1995).

It was noted in Section 5.1 that snow cover has a distinctive variation of emissivity with frequency, and several algorithms have been designed to exploit this fact using multifrequency data. For example, the approach of Grody and Basist (Grody 1991; Basist and Grody 1994; Grody and Basist 1996) performs a series of tests on data from the SSM/I instrument to identify scattering materials and then to eliminate those that are not snow cover. The tests are defined as follows:*

$$(T_{22V} - T_{85V} > 0) \text{ OR } (T_{19V} - T_{37V} > 0) \qquad (5.8)$$

identifies a scattering medium,

$$(T_{22V} > 257) \text{ OR } (T_{22V} - 0.49T_{85V} > 165) \qquad (5.9)$$

indicates precipitation,

$$(T_{19V} - T_{19H} > 17) \text{ AND } (T_{19V} - T_{37V} < 10) \text{ AND } (T_{37V} - T_{85V} < 10) \quad (5.10)$$

indicates cold desert (such as Central Iran, the Gobi Desert, and the Tibetan Plateau), and

$$((T_{22V} - T_{85V} < 8) \text{ OR } (T_{19V} - T_{37V} < 8)) \text{ AND } (T_{19V} - T_{19H} > 7) \qquad (5.11)$$

indicates areas of frozen ground. Removal of areas of precipitation remains a problem if passive microwave data are used alone, since snow cover and precipitation can give similar signatures (Standley and Barrett 1999; Negri et al. 1995; Bauer and Grody 1995; Standley and Barrett 1995). Synergies with thermal infrared have been explored as a way of resolving this ambiguity. These approaches have included the use of the SSM/T instrument (Bauer and Grody 1995) and the higher-resolution OLS instrument (Standley and Barrett 1999), both carried by the DMSP satellites that also carry the SSM/I passive microwave radiometers.

Algorithms suitable for defining the extent of snow cover can also be derived by modifying algorithms for snow depth (Section 5.3). For example, Hall et al. (2002) adopt a modification of the snow depth algorithm of Chang, Foster, and Hall (1987), since they find this to be more reliable than the algorithm of Grody and Basist (1996). The snow depth (in mm) is estimated from

$$SD = 15.9(T_{19H} - T_{37H}) \qquad (5.12)$$

The pixel is mapped as snow if

$$SD > 80 \text{ AND } T_{37V} < 250 \text{ AND } T_{37H} < 240 \qquad (5.13)$$

This formula, or others like it intended for global snow mapping, requires some modification for use at high altitude, for example, for mapping snow cover in

---

*In these and subsequent equations, the notation T_{fp} means the brightness temperature in kelvin measured at a frequency of f GHz in polarization state p.

the Himalayas (Saraf et al. 1999). Here the smaller optical thickness of the atmosphere changes the coefficients somewhat. The following has been proposed for SMMR data (Saraf et al. 1999):

$$SD = 20(T_{18H} - T_{37H}) - 80 \qquad (5.14)$$

where the snow depth SD is again expressed in millimeters.

A different approach is due to Mätzler (1994). This involves estimating the land surface temperature from passive microwave data, using this temperature to deduce the effective emissivities at different frequencies and polarizations, and then applying a set of rules to these emissivities to determine whether snow is present. In the form of the procedure developed by (Hiltbrunner and Mätzler 1997), the steps can be represented as follows. First, the land temperature (in kelvin) is estimated as

$$T = \frac{1.95T_{19V} - 0.95T_{19H}}{0.95} \qquad (5.15)$$

Effective emissivities are calculated as ratios of observed brightness temperatures to the land temperature, and the following linear combination of emissivities is defined from them:

$$C = \varepsilon_{19V} - \varepsilon_{19H} + \varepsilon_{37V} - \varepsilon_{37H} + 3(\varepsilon_{19V} - \varepsilon_{37V}) \qquad (5.16)$$

An area is classified as snow-covered if

$$C > 0.14 \text{ AND } T < 293 \text{ K} \qquad (5.17)$$

This procedure appears to work well except in areas of forest cover.

A review of rival passive microwave algorithms by Armstrong and Brodzik (2002) concludes that overall snow extent is best estimated by the use of horizontally polarized channels (Chang, Foster, and Hall 1987), while vertical-polarization algorithms (Goodison 1989) tend to overestimate snow cover on frozen ground or desert. Passive microwave algorithms generally tend to underestimate snow extent, with the largest discrepancies occurring in autumn and the smallest in spring. This is most likely to be a consequence of thin snow packs, less than about 3 cm in depth (Chang, Foster, and Hall 1987; Hall et al. 2002). Sensitivity to these thin snow packs can be improved by including data from the SSM/I channel at 85 GHz (Nagler 1991; Grody and Basist 1996) but this tends to overestimate the total snow extent.

As with intermediate-scale mapping, the problem of complex (heterogeneous) terrain is significant for global-scale mapping of snow cover. An analysis of this problem has been carried out by Pivot, Kergomard, and Duguay (2002) using a principal components analysis (Section 3.3.3) of time series data to identify spatiotemporal variations. A particular difficulty occurs in areas "contaminated" by trees, since the tree canopy projects above the

snow cover and also emits microwave radiation. An approach to this problem
has been developed by Tait (1998) using a land cover classification.

In practice, global snow cover mapping is usually based on a number of
sources of data. An automated procedure using GOES, AVHRR, and SSM/I
data has been developed for North America (Romanov, Gutman, and Csiszar
2000). It provides a resolution of 0.04° (around 4 km) and is estimated to
coincide with manual classification of multisource data about 85% of the time.
The NOAA/NESDIS northern hemisphere weekly snow charts (which have
in fact been produced daily since 1999) are composite products based on
the manual analysis of AVHRR, GOES, and other optical satellite data
(Robinson, Dewey, and Heim 1993; Ramsay 1998) (http://www.ssd.noaa.gov).
The National Operational Hydrologic Remote Sensing Center (NOHRSC)
produces weekly snow charts for North America with a resolution of 1.1 km
(http://www.nohrsc.nws.gov) using AVHRR data (König, Winther, and
Isaksson 2001). A similar procedure is used by the Norwegian Water
Resources and Environment Directorate (NVE) (Andersen 1982; König,
Winther, and Isaksson 2001). Since 1999, data from the MODIS instrument on
board the Terra satellite have been used by NESDIS to provide daily and
8-day composite snow cover data at a resolution of 500 m or 1 km. Figure 5.5
shows the total snow extent in the northern hemisphere from 1979 to 2003,

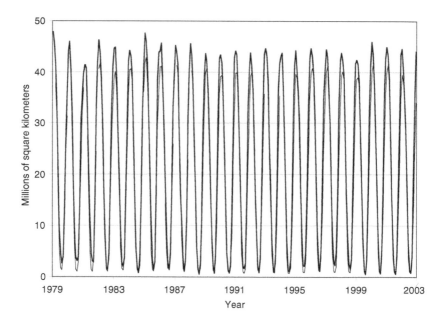

FIGURE 5.5 Northern hemisphere snow cover (excluding Greenland), 1979 to 2003,
measured using VIR (heavy curve) and passive microwave (light curve) imagery. (Data
supplied by R.L. Armstrong and M.J. Brodzik, National Snow and Ice Data Center,
Boulder, Colorado (http://nsidc.org/sotc/snow_extent.html)).

estimated from satellite data. The data show a trend of about −0.4% per annum in the mean snow cover, although this is not uniform over the northern hemisphere, with some regions showing trends in the opposite direction. Most of the change is attributable to decreases in the snow cover in spring and summer since the mid-1980s, while autumn and winter snow cover have exhibited no significant trend (Robinson 1997, 1999). Comparison with historical data from *in situ* measurements suggests that this decrease followed a general increase in snow cover extent during the 20th century, mainly in winter (Brown 2000). A more recent analysis of the NOAA weekly snow cover data suggests that a decrease in snow cover extent during the 1980s was followed by an increase during the 1990s (Brown and Braaten 1998).

5.3 SNOW WATER EQUIVALENT AND SNOW DEPTH

5.3.1 SMALL AND INTERMEDIATE SCALES

Various techniques are available for estimating snow depth and snow water equivalent (SWE) over local to regional areas. Snow depth can be determined from airborne digital stereophotogrammetry (Cline 1993; Smith, Cooper, and Chapman 1967). The idea is to measure the elevation of the snow-free surface and then the snow-covered surface. The difficulty is in identifying conjugate (matching) points in the stereo images from a comparatively featureless material like snow. A possible extension of this idea is the use of airborne laser profiling.

With the exception of photogrammetric methods, snow depth cannot be determined directly from VIR imagery (Stähli, Schaper, and Papritz 2002), although indirect methods are possible. One approach is to use a *snow depletion model* (Cline, Bales, and Dozier 1998), which relates the areal coverage of snow to its mean depth and is, as its name implies, designed primarily for relating snow extent to runoff during the melt season. The accuracy of this approach can be improved by establishing a regression relation with altitude and using a DEM (König, Winther, and Isaksson 2001). Attempts to include vegetation cover as a second explanatory variable in vegetated areas have been proposed (Forsythe 1999; Stähli, Schaper, and Papritz 2002), and they provide some further improvement in accuracy except in forested areas (Stähli, Schaper, and Papritz 2002).

SWE can be estimated from airborne measurements using gamma-ray surveying. The naturally occurring gamma emission is attenuated by the mass of overlying water, whether ice or snow, and can be used to calculate the SWE up to about 35 cm (Dahl and Ødegaard 1970; Carroll and Carroll 1989). This procedure is carried out regularly over the Great Plains of North America.

It has been suggested that polarimetric SAR can also be used to measure snow depth, using the difference between the VV- and HH-polarized backscattering coefficients (Shi, Dozier, and Davis 1990; Shi and Dozier 1995).

A method for measuring dry-snow SWE using InSAR has also recently been proposed by Guneriussen et al. (2000).

5.3.2 GLOBAL SCALE

Although recent research has indicated that snow depth can be measured by radar altimetery (Papa et al. 2002), passive microwave radiometry remains the only spaceborne technique with proven ability to extract SWE data (Walker and Silis 2002). The physical basis for the technique is the dependence of the amount of volume scattering on the quantity of snow in the pack, which reduces the brightness temperature of the radiation emitted from the underlying ground. This is only possible for dry snow, where volume scattering is the dominant mechanism. Much research activity has been devoted to extracting SWE from passive microwave data (Chang, Foster, and Hall 1990; Goodison and Walker 1995; Tait 1998; Pulliainen and Hallikainen 2001; Chang, Foster, and Hall 1987; Hallikainen and Jolma 1992; Grody and Basist 1996), and a number of algorithms have been developed. These generally exploit the difference in emissivity between two frequencies, typically 18 or 19 GHz and 37 GHz. For example, the Meteorological Service of Canada (MSC) has developed algorithms for central Canada based on differences between the 19V and 37V SSM/I channels. The general form of the retrieval algorithm is

$$\text{SWE} = a - b(T_{37\text{V}} - T_{18\text{V}}) \tag{5.18}$$

where SWE is the snow water equivalent in mm and a and b are empirical coefficients. The nature of the underlying land cover can have a significant effect on the estimated SWE (Kurvonen and Hallikainen 1997), and in the MSC algorithm different values of the coefficients are used for different land cover types (open, deciduous, coniferous, and sparse forest). Over open prairie, the values are $a = -20.7$, $b = 2.59$ (Walker and Silis 2002). Different coefficients are used if the 18 GHz channel is replaced by one at 19 GHz. In this case, Josberger et al. (1998) propose values of $a = -0.7$, $b = 5.14$. The results of this and earlier algorithms have been used operationally since 1988 (Derksen et al. 2002; Goodison and Walker 1995), and have been shown to be generally accurate to within 10 or 20 mm SWE of values measured *in situ*, although serious underestimates (30 mm or more) can occur in areas where many lakes are present or in areas of exceptionally high SWE (Derksen et al. 2002; Walker and Silis 2002; De Sève et al. 1997). The effect of grain size is also significant: larger grains reduce the effective emissivity of the snow pack and hence the observed brightness temperature (Chang et al. 1981; Hallikainen 1986) (Figure 5.6). The effect of snow metamorphism (Section 4.2.1) on emissivity is complicated and introduces errors to existing algorithms for snow properties as the winter progresses (Rosenfeld and Grody 2000). However, if there is significant variation of the snow grain size within the snow pack, substantial

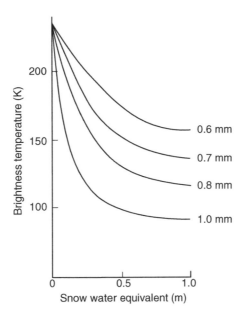

FIGURE 5.6 Effect of grain size on microwave brightness temperature (37 GHz, vertical polarization) of snow. (Data from Chang et al. (1981) presented by Armstrong et al. (1993). Figure reprinted from *Annals of Glaciology* with permission of the International Glaciological Society.)

underestimates in the snow depth (of the order of a factor of 2) can be incurred using a simple model that assumes constant grain size (Mognard and Josberger 2002; Sturm, Grenfell, and Perovich 1993). This situation typically occurs in very cold snow packs where the temperature gradient can exceed about 20 K m^{-1} (Armstrong 1985), such as the Great Plains of the United States, Canada, Alaska, and Siberia. It can be corrected for by using air temperature data to estimate the temperature gradient in the snow pack (Josberger and Mognard 1998; Mognard and Josberger 2002).

In forest-covered areas, a correction has been proposed by Chang and Chiu (1991):

$$\text{SWE} = a(T_{19H} - T_{37H}) + f(T'_{19H} - T'_{37H}) \qquad (5.19)$$

where T' is the brightness temperature measured for a snow-free forest-covered pixel and f is the fraction of a pixel covered by forest. The coefficient a is determined empirically.

An alternative algorithm has been proposed by Hallikainen (1984), which involves calculating the difference ΔT in brightness temperatures between a winter and an autumn image of the same area:

$$\Delta T = (T_{18V} - T_{37V})_{\text{winter}} - (T_{18V} - T_{37V})_{\text{autumn}} \qquad (5.20)$$

The SWE is then calculated as

$$\text{SWE} = a\Delta T - b \qquad (5.21)$$

where the coefficients a and b were determined to have the values 10.1 and 98.0 for southern Finland and 8.7 and 108.0 for northern Finland. The temporal differencing reduces the effect of variations in land cover.

More recently, efforts have been directed toward the development of neural network methods of retrieving SWE (Chang and Tsang 1992; Lure et al. 1992; Tsang et al. 1992; Davis et al. 1993; Wilson et al. 1999) from passive microwave data.

An interesting new idea for the determination of the distribution of snow over large scales is the use of gravity measurements by satellite (Mognard 2003).

5.4   SNOW MELT AND RUNOFF MODELING

Remotely sensed data play an important role in monitoring the progress of snow melt and modeling the quantity of water produced as runoff. The simplest model of snow melt uses a "degree–day coefficient" of typically 0.5 cm, i.e., one degree–day of thaw, without rainfall, can melt 0.5 cm of water from the snow pack. More sophisticated models include the effect of precipitation, and also the time lag between melting of snow and the appearance of the resulting meltwater at the gauging station (Martinec, Rango, and Major 1983; Hall and Martinec 1985). The snowmelt runoff model (SRM) of (Martinec, Rango, and Major 1983) is widely used, for example in the ASCAS system. It can be written as

$$Q_{n+1} = Af(1 - k_{n+1})(c_{s,n}\alpha_n[T_n^+ + \Delta T_n] + c_{r,n}P_n) + k_{n+1}Q_n \qquad (5.22)$$

where Q_n is the modeled discharge (volume per unit time) on day n, A is the area of the drainage basin or an altitudinal zone of it, k is the recession coefficient (this describes how the runoff varies with time if there is no contribution from melting or precipitation), c is a runoff coefficient with subscript "s" for snowmelt, "r" for rainfall, T^+ is the number of degree–days of thaw, adjusted by ΔT to allow for the lapse rate between the place where the temperature is measured and the drainage basin or zone, P is the rainfall contribution to runoff, and f is a factor to convert units. Simpler variants of this model describe the runoff as a linear combination of precipitation, temperature and fractional snow cover (Wang and Li 2001; Schaper, Seidel, and Martinec 2000).

It is clear from equation (5.22) that the application of the SRM requires both the area of snow cover, which can be obtained from remotely sensed imagery (Seidel and Martinec 1992; Shashi Kumar et al. 1992; Swamy and Brivio 1996), and ancillary data such as temperature, precipitation, and runoff, which cannot be obtained in this way. The SRM is essentially a form of geographic information system in which data from different sources are fused. The ability to distinguish between frozen and melting snow cover can enhance the performance of the model. This can be provided from VIR or especially SAR imagery (Koskinen, Pulliainen, and Hallikainen 1997; Maxfield 1994; Nagler and Rott 1997; Running et al. 1999) or microwave scatterometry.

5.5 PHYSICAL PROPERTIES OF THE SNOW PACK

5.5.1 REFLECTANCE AND ALBEDO

Satellite monitoring of albedo is well established (König, Winther, and Isaksson 2001). The process can be summarized as follows, based on König, Winther, and Isaksson (2001): first the *planetary reflectance* is calculated as the ratio of the radiance measured at the satellite to the above-atmosphere (exoatmospheric) solar irradiance in the spectral band in question (the latter can be calculated from the known properties of solar radiation and the distance from the Sun to the Earth). If it is assumed that the surface is a Lambertian reflector, the formula for the planetary reflectance R is

$$R = \frac{\pi L}{E_{sun} \cos \theta} \qquad (5.23)$$

(e.g., Koelemeijer, Oerlemans, and Tjemkes 1993), where L is the observed radiance, E_{sun} is the solar irradiance, and θ is the sun's zenith angle. This reflectance is corrected for atmospheric transmission effects using a radiative transfer model, for example, the "5S" code (Tanré et al. 1990). Failure to correct for atmospheric effects can result in large errors in the reflectance (Hall et al. 1989).

For energy-balance calculations the broad-band reflectance, i.e., the reflectance averaged over the effective range of the solar spectrum (say 0.3 to 2.5 μm) is needed, whereas satellite sensors generally provide narrow-band data. Various approaches can be adopted, including a linear or quadratic combination of the narrow-band reflectances (Duguay and LeDrew 1992; Knap, Reijmer, and Oerlemans 1999; Li and Leighton 1992; Knap and Oerlemans 1996; Jacobsen, Carstensen, and Kamper 1993), interpolation of the reflectance spectrum in the visible part of the spectrum (where the reflectance does not vary much) (Hall et al. 1989), and calculations based on theoretical models (e.g., Choudhury and Chang 1979) of snow reflectance as a function of grain size (Section 5.5.2). To convert the reflectance to hemispherical (diffuse) albedo,[*] it is necessary

---

[*]This can be defined as the albedo averaged over all possible incidence directions. It is also equal to the ratio of the total scattered power to the total incident power when the latter is distributed isotropically.

also to know the bidirectional reflectance distribution function of the snow surface. This is often assumed to be Lambertian, although as discussed in Section 4.2.4, snow exhibits a significant anisotropy.

In areas of appreciable topographic relief, it is important to perform topographic normalization to reduce the effects of slope orientation, shadowing, and variable atmospheric transmission (Xin, Koike, and Guodong 2002; Dozier 1984; Dozier and Marks 1987). Various methods have been proposed to solve the problem of topographic normalization, as summarized by Xin, Koike, and Guodong (2002). These include band ratioing (Holben and Justice 1980), application of the "cosine law" or the Minnaert model* (Eyton 1989; Colby 1991), and deterministic models that model both solar irradiation and atmospheric correction through the use of a digital elevation model and the assumption of Lambertian scattering from the snow surface (Proy, Tanré, and Deschamps 1989; Dozier and Frew 1990). Improved models that directly include the effect of diffuse radiation (skylight) have been used by Xin, Koike, and Guodong (2002) and Scherer and Brun (1997).

A preliminary albedo algorithm for MODIS data has been developed by Klein and Hall (1999), with products available from the U.S. National Snow and Ice Data Center (http://nsidc.org). Strictly speaking, the quantity that is estimated is the directional hemispheric reflectance, since the algorithm assumes that the incident radiation is collimated. The procedure is outlined by Klein and Stroeve (2002). Briefly, it consists of the following stages: (1) calibration of the sensors to determine the top-of-atmosphere (TOA) reflectances in the narrow MODIS wavebands; (2) correction for atmospheric effects (Vermote and Vermeulen 1999) and the effects of anisotropic reflectance, using a DEM to obtain surface slope and the discrete-ordinate radiative transfer model (DISORT) of Stamnes et al. (1988); (3) narrow-band to broad-band conversion, i.e., derivation of the spectrally integrated albedo.

5.5.2   GRAIN SIZE

Determination of grain size is important, since it gives information about the thermal history of the snow pack (Tanikawa, Aoki, and Nishio 2002). The grain size of dry snow affects the reflectance in the near-infrared region (Dozier 1984, 1989; Wiscombe and Warren 1980), and this can be used to estimate the grain size from satellite imagery (e.g., Bourdelles and Fily 1993; Dozier, Schneider, and McGinnis 1981; Fily et al. 1997; Fily, Dedieu, and Durand 1999; Shi and Dozier 1994). The most reliable method appears to be based on

---

*The Minnaert model assumes that the bidirectional reflectance distribution function of the surface is proportional to $(\cos\theta_0 \cos\theta_1)^{a-1}$ where θ_0 and θ_1 are, respectively, the angles between the incident and reflected radiation and the surface normal, and a is a constant. The cosine law corresponds to the case of Lambertian scattering, when $a = 1$.

wavelengths corresponding to Landsat TM (or ETM+) bands 5 and 7 (1.55 to 1.75 μm and 2.08 to 2.35 μm, respectively). At shorter wavelengths, penetration of radiation into the snow pack introduces complications. The data must be corrected for atmospheric propagation and for the viewing geometry. Various authors have described the retrieval of grain size from remotely sensed data by inverting scattering models (e.g., Bourdelles and Fily 1993; Fily et al. 1997) using Landsat TM data and (Nolin and Dozier 1993, 2000) using airborne AVIRIS data. However, these studies did not include the effect of impurities in the snow, and these can significantly alter the albedo, especially in the near-infrared part of the spectrum (Warren and Wiscombe 1980). Furthermore, most attempts to retrieve snow grain size from reflectance measurements assume isotropic scattering from the snow, and this can be a poor assumption (Grenfell, Warren, and Mullen 1994). A recent study that includes both effects is described by Tanikawa, Aoki, and Nishio (2002). This uses airborne imagery at wavelengths of 0.545, 1.24, 1.64, and 2.23 μm to infer grain size and the concentration of impurities.

Snow grain size also influences the microwave properties of the snow pack. As was noted in Section 5.3.2, the microwave emissivity is affected by grain size, in particular the spectral gradient of brightness temperature at the low-frequency end of the microwave spectrum (e.g., between 6 and 18 GHz) (Sherjal et al. 1998; Surdyk and Fily 1993). Radar measurements can also be used to estimate snow pack parameters, including grain size. In general the dependence is complicated, being contributed to by surface and volume scattering as well as the underlying terrain (Sections 4.2.7 and 5.5.4). The use of multifrequency and/or multipolarimetric radar will be helpful in this regard, since the effects of the physical parameters vary differently with frequency (Kuga et al. 1991).

5.5.3 TEMPERATURE

Snow surface temperature can be derived from thermal infrared measurements. Over the range 10.5 to 12.5 μm the emissivity of snow is close to unity* (Warren 1982) (Section 4.2.5) so that Landsat band 6 or equivalent provides a good means of measuring temperature. Atmospheric propagation introduces errors of up to about 10 K but these can be corrected by modeling or by calibration against surface data, since the atmospheric effects do not vary rapidly with position (Orheim and Lucchitta 1988; Pattyn and Decleir 1993). The dual viewing direction of the ERS ATSR instrument allows atmospheric correction to be estimated directly from the data (Bamber and Harris 1994; Stroeve, Haefliger, and Steffen 1996).

---

*Salisbury, D'Aria, and Wald (1994) have, however, reported dependence of emissivity on grain size, density, and liquid water content.

5.5.4 RADAR PROPERTIES

By radar properties we mean those properties of the snow pack that control
the microwave backscattering coefficient. As has been noted in Section 4.2.7,
several factors are involved. If the snow is dry and deep, the principal
mechanism is volume scattering from the snow pack, and the most important
variables are the density and grain size. For shallower dry snow, the depth of
the snow pack and the contribution from the underlying terrain are also
significant. In the case of wet snow, the principal mechanism is surface
scattering and the most important variables are the dielectric constant of the
snow (controlled by its density and liquid water content), its surface roughness
properties, and the local incidence angle. In general, then, a radar image that
measures a single variable (e.g., the backscattering coefficient at a single
frequency and polarization state) will not be able to disentangle the effects of
these variables and can merely define ranges of consistent values for them.
Nevertheless, several useful quantitative studies have been performed. As an
example, a detailed investigation of the dependence of SAR backscatter with
snow properties was presented by Guneriussen (1997) using ERS data from the
Kvikne area in southern Norway. He used a Kirchhoff stationary phase model
for surface scattering plus a Rayleigh model for volume scattering (Section
4.2.6), fitting the values of RMS slope and mean dielectric constant.
Backscatter data were first corrected for local geometry using a DEM. Two
approaches were adopted: (1) physical modeling using the Kirchhoff/Rayleigh
model; (2) empirical correction, in which the backscatter coefficient is assumed
to vary in a simple manner with local incidence angle, for example, using the
Muhleman model

$$\sigma^0 \propto \cos \theta \tag{5.24}$$

or the modified Muhleman model (Stiles and Ulaby 1980a):

$$\sigma^0 \propto \frac{\cos \theta}{(\sin \theta + 0.1 \cos \theta)^3} \tag{5.25}$$

6 Remote Sensing of Sea Ice

6.1 INTRODUCTION

Like snow cover on land, sea ice is an easy phenomenon to recognize in remotely sensed imagery. In the visible and near-infrared region of the electromagnetic spectrum it exhibits a markedly higher albedo than the background — open water — against which it must be recognized (see Color Figure 2.4 following page 108). VIR observations from space, with a latitudinal coverage large enough to view most of the polar regions, began in the mid-1960s with the launch of TIROS-9 and the early NIMBUS and ESSA satellites, although the first spaceborne instrument with a spatial resolution of around 1 km was not deployed until the early 1970s. This was NOAA-2, which carried, among other instruments, the VHRR (Very High Resolution Radiometer), the predecessor of the AVHRR. The value of the AVHRR was recognized early, and applied to the monitoring of sea ice (Hufford 1981) although its susceptibility to cloud cover, and difficulty in recognizing thin ice, have continued to be obstacles to the use of such data (Steffen et al. 1993).

The microwave emissivity of sea ice also differs significantly from that of open water, so that the two materials are easy to distinguish in passive microwave imagery (Parkinson 2002; Parkinson and Gloersen 1993) (Figure 6.1). Passive microwave radiometry has been used for monitoring sea ice from space since the first spaceborne passive microwave radiometer, the ESMR (Electrically Scanned Microwave Radiometer), was deployed in 1972. Thus we can effectively date the start of spaceborne monitoring of sea ice to the early 1970s.

The requirements for remote sensing of sea ice do not, of course, end with the ability to detect the presence of sea ice. We will probably also want to determine the type of ice (first-year or multiyear, perhaps with further subdivision of type), its dynamic behavior, its thickness, the size distribution of floes and leads, and perhaps other physical parameters. We also need to recognize that wide-swath measurements that are good for obtaining synoptic measurements over large areas will probably not have sufficient spatial resolution to resolve individual floes, and this introduces the concept of *ice concentration* as a spatially averaged proportion of ice cover.

6.2 ICE EXTENT AND CONCENTRATION

Sea ice extent can readily and straightforwardly be determined from cloud-free VIR imagery on the basis of the large difference in reflectance between

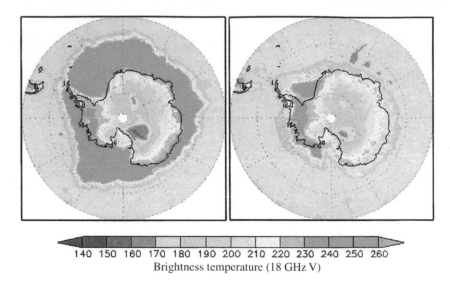

140 150 160 170 180 190 200 210 220 230 240 250 260
Brightness temperature (18 GHz V)

FIGURE 6.1 (See Color Figure 6.1 following page 108) Brightness temperature over
Antarctica measured with the Oceansat-1 Multichannel Scanning Microwave Radio-
meter (MSMR). (Left) 12 to 18 September 1999; (right) 13 to 19 March 2000.
(Figure adapted from Dash et al. (2001), with permission of Taylor & Francis plc.

ice and open water (Zibordi and Meloni 1991; Zibordi and van Woert 1993)
(Figure 6.2). The contrast is very much reduced in the case of thin ice (visible
in the eastern part of the Kara Sea in Figure 6.2), but otherwise simple
thresholding, preceded by masking-out of cloud-covered areas, is sufficient.
In practice, much analysis of VIR imagery has been performed manually,
but Williams et al. (2002) describe an automated rule-based procedure,
ICEMAPPER, based on AVHRR imagery and a land/sea mask. The
procedure is summarized in Figure 6.3, and is augmented by rules to determine
the probable ice cover below thin cloud. Once areas of sea ice have been
determined, the concentration is estimated by linear interpolation of the albedo
between the values for open water and for 100% ice cover.

Sea ice extent is currently calculated on a daily basis, at a resolution of
1 km, from MODIS imagery* (Riggs, Hall, and Ackerman 1999).

The principal disadvantage of the use of VIR data is its susceptibility
to cloud cover. Passive microwave imagery is not affected by cloud cover, so
offers a valuable alternative as a technique for identifying sea ice (Comiso et al.
1997). As was noted in Section 6.1, the coarse spatial resolution of passive
microwave imagery in relation to the typical size of ice floes implies that the

---

*http://modis-snow-ice.gsfc.nasa.gov/sea.html.

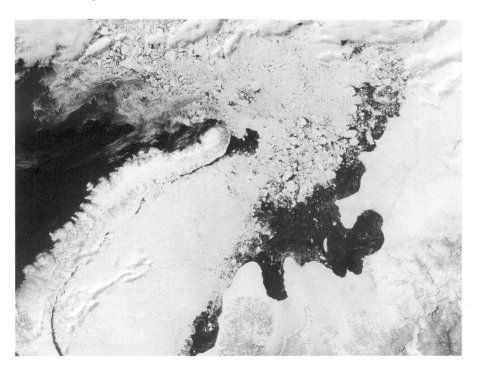

FIGURE 6.2 MODIS image of Novaya Zemlya and Kara Sea, 10 June 2001. The Barents Sea (west of Novaya Zemlya) is largely free of ice, while the Kara Sea has an extensive ice cover. A few clouds are visible in the image.

approach must necessarily be a statistical one, determining the ice concentration rather than delineating floes.

The first spaceborne passive microwave radiometer, ESMR, provided a single channel of data (19.4 GHz, H-polarized). Ice concentrations were retrieved using the following formula, which is based on linear interpolation:

$$C = \frac{T_b - \varepsilon_w T_w}{\varepsilon_i T_i - \varepsilon_w T_w} \qquad (6.1)$$

(Parkinson and Gloersen 1993; Steffen et al. 1992) where C is the ice concentration, T_b is the observed brightness temperature, ε_w and ε_i are, respectively, the emissivities of open water and of sea ice, and T_w and T_i are the physical temperatures of water and ice. The principal difficulty in applying this formula is that ε_i varies substantially according to the type and condition of the ice. Multichannel instruments, such as the SMMR (Scanning Multichannel Microwave Radiometer) onward, substantially resolve the ambiguities. The NASA Team algorithm (Cavalieri, Gloersen, and Campbell 1984; Gloersen and

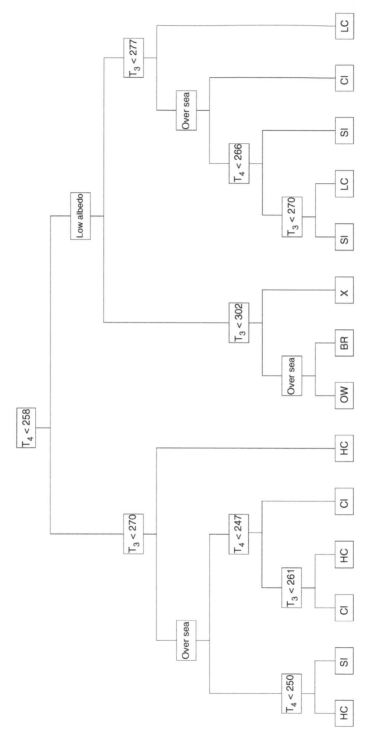

FIGURE 6.3 ICEMAPPER decision tree (adapted from Williams et al. 2002). HC = high cloud, LC = low cloud, CI = continental ice, SI = sea ice, OW = open water, BR = bare rock, X = interference.

Cavalieri 1986; Steffen et al. 1992) involves the use of the *polarization ratio* PR and the *gradient ratio* GR,[*] defined as

$$PR = \frac{T_{18V} - T_{18H}}{T_{18V} + T_{18H}} \tag{6.2}$$

and

$$GR = \frac{T_{37V} - T_{18V}}{T_{37V} + T_{18V}} \tag{6.3}$$

where T_{18V} denotes the vertically polarized brightness temperature at 18 GHz, and similarly. Concentrations of particular ice types are then calculated from formulae of the form

$$C = \frac{a + bPR + cGR + dPR \cdot GR}{e + fPR + gGR + hPR \cdot GR} \tag{6.4}$$

where the coefficients a to f are determined empirically and differ for different ice types, and indeed between the northern and southern hemispheres. They can be tuned locally, using *tie points* (data values representing open water, 100% first-year ice and 100% multiyear ice, or hemispherically using a statistical analysis of a whole year's data.

The formulae of equation (6.4) define an approximately triangular region (in fact, the edges are somewhat curved) in a plot of GR against PR, illustrated schematically in Figure 6.4. The points labeled OW, FY, and MY correspond to pure open water, first-year ice, and multiyear ice, respectively. The main source of potential error in this approach is weather-dependent effects (attenuation and emission by liquid water in clouds, and roughening of water surface by wind). Most of these can be eliminated by recognizing that only points lying within the triangular region are physically meaningful (Gloersen and Cavalieri 1986).

There are other algorithms for the retrieval of ice concentration from passive microwave data, reviewed by Steffen et al. (1992). These include the AES-York, FNOC, NORSEX (Svendsen et al. 1983), University of Massachusetts, and Bootstrap algorithms (Figure 6.5). These algorithms have different regional and hemispheric characteristics, as indicated by Figure 6.4. New approaches based on artificial intelligence methods, such as knowledge-based and neural-network classification, show some promise.

The intrinsic accuracy of the total ice concentration retrieved from passive microwave data is about ±10% when the surface is dry (Steffen and Schweiger 1991; Cavalieri et al. 1991), though the error is much increased by the presence of melt features, and the algorithms are generally unreliable in

---

[*]The great advantage of using the ratios PR and GR is that it eliminates the necessity to estimate the physical temperatures of ice and water.

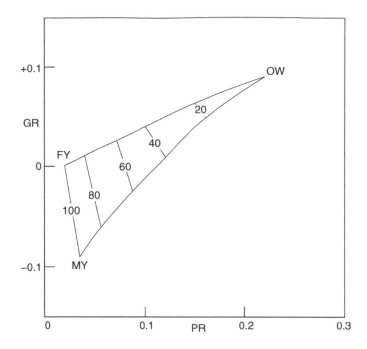

FIGURE 6.4 Schematic representation of the NASA Team algorithm. The position of a point in the diagram indicates the total ice concentration (shown in percent) and the relative proportions of first-year (FY) and multiyear (MY) ice.

summer. Ice with a significant coverage of melt ponds appears as lower-concentration ice (El Naggar, Garrity, and Ramseier 1998). The presence of wet-snow cover can also introduce ambiguities (Garrity 1992). Thin ice (less than about 0.3 m) can also be a problem — it can appear as 30% open water (Grenfell et al. 1992). Concentrations of individual ice types are less reliable than the total ice concentration. In particular, passive microwave algorithms tend to underestimate the fraction of multiyear ice significantly (Comiso 1990).

Sea ice concentration data from passive microwave imagery are archived by the U.S. National Snow and Ice Data Center (http://nsidc.org), gridded at a spacing of 25 km (although the spatial resolution of the data is approximately 50 km). The datasets are derived from the SMMR instrument (1978 to 1987) and from the SSM/I from 1987 onwards, and were merged by Cavalieri et al. (1999) using the NASA Team algorithm (Gloersen et al. 1992). The latitudinal coverage of the two instruments is different, being set by the swath width and the orbital parameters of the satellites on which they are carried. SMMR data extend to a latitude of 84.6°, while SSM/I extend to 87.6°.

Other remote sensing techniques can be used to infer sea ice concentration. These include the use of synthetic aperture radar (Burns et al. 1987), microwave scatterometry (Grandell, Johannessen, and Hallikainen 1999), and radar

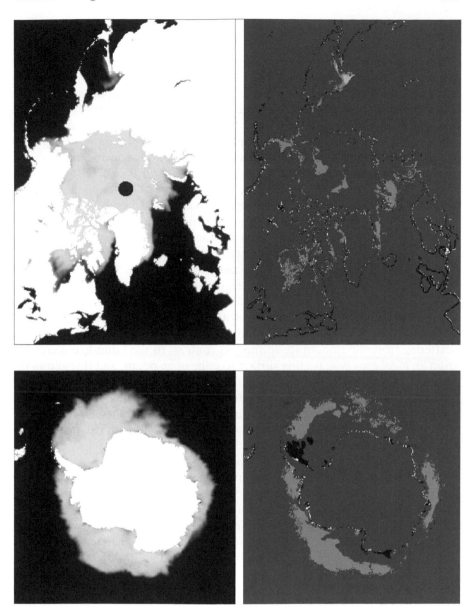

FIGURE 6.5 (Left) Average ice concentrations in the northern hemisphere, December 2003, and in the southern hemisphere, June 2003, calculated from SSM/I data using the NASA Team algorithm. Progressively lighter shades of gray denote increasing total ice concentration. (Right) Ice concentrations from the bootstrap algorithm minus ice concentrations from the NASA Team algorithm. Black: −6% and less; dark gray: −6 to +6%; mid-gray: +6 to 18%; light gray: +18 to +30%; white: +30% and above.

altimetry (Fetterer et al. 1992), three microwave techniques that, like passive microwave radiometry, are also largely unaffected by cloud cover and entirely independent of daylight. The development of synthetic aperture radar (SAR) has generally been in the direction of higher spatial resolution at the expense of swath width, and this has militated against its use for synoptic studies of sea ice (although its utility for more detailed investigations is conversely increased, and discussed below). More recently, beginning with the launch of Radarsat in 1995, spaceborne SARs with multiple modes of observation have begun to be deployed. The SARs carried by Radarsat and Envisat include wide-swath modes of the order of 500 km. A similarly wide swath, valuable for synoptic studies, can be provided by a spaceborne SLR system.* The acronym stands for "side-looking radar" and the technology is essentially the predecessor to SAR. It employs the same side-looking imaging geometry but does not use the signal processing employed by SAR to increase the along-track resolution (Section 2.9). Spaceborne SLR systems have principally been developed in the former Soviet Union. Figure 6.6 shows a mosaic of four strips of SLR imagery from the Kara Sea, together with the coverage of coincident narrow-swath ERS SAR images.

In the case of radar altimetry, the waveform (time-variation) of the signal reflected from sea ice is usually sharply peaked because of significant specular scattering from the smooth ice surface, in contrast to the ramp-like waveform from an incoherently scattering surface like the open ocean, and this allows the presence of sea ice to be inferred (Laxon 1989). However, the mechanisms are not fully understood and there are also technical difficulties associated with the problem of tracking the return from a very rough surface (Fetterer et al. 1992). One approach to the analysis of radar altimetry data is due to Chase and Holyer (1990), who applied a linear unmixing model to the waveforms. This involves identifying the waveforms representative of particular ice types (including open water), representing the observed waveform as a linear combination of these "end-members" and deducing the proportions of each that are present in an observed waveform. They found agreement to within 5% of the concentrations calculated from passive microwave data. Alternative approaches include the use of the measured backscattering coefficient (Ulander 1991) or the total power in the waveform (Drinkwater 1991).

As was remarked in Section 6.1, sea ice has been monitored from space since the early 1970s. Have significant long-term trends been discovered as a result? Early studies using passive microwave data were suggestive of such trends but did not have a long enough time span to cover the El Niño, La Niña, and North Atlantic Oscillation phenomena. More recently, analyses have been performed for periods of about 20 years. Analyzing northern hemisphere data for the period 1981 to 2000, Comiso (2002) reports a mean trend in the ice cover of $-24,600 \text{ km}^2 \text{ a}^{-1}$, equivalent to an annual decrease of $0.2 \pm 0.03\%$ of

---

*Other names are also in use, including Real Aperture Radar (RAR) and Side-Looking Airborne Radar (SLAR).

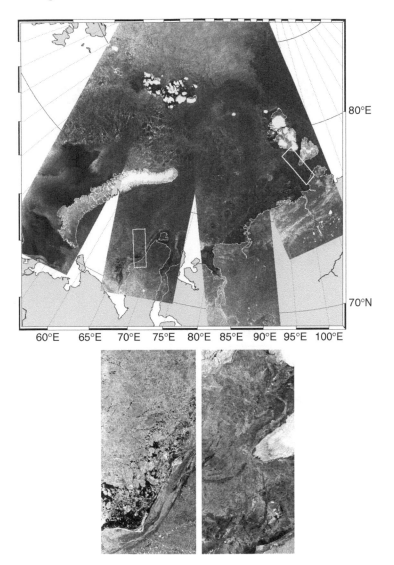

FIGURE 6.6 (Above) Digital mosaic of four SLR images of the Kara Sea obtained from the OKEAN-N7 satellite, 16–21 May 1996. (Copyright NPO Planeta, Moscow, department of thematic processing, 1996.) (Below) Simultaneous ERS SAR images from the areas indicated by the white boxes. (Copyright ESA/Tromsø Satellite Station, 1996. Images reproduced by courtesy of Nansen Environmental and Remote Sensing Center, Bergen.)

the mean ice extent. Parkinson et al. (1999) report a similar but slightly higher rate of decrease (0.28% per annum) for the period 1978 to 1996. The actual area of ice, found by integrating the concentration with respect to area, appears to be decreasing slightly faster because the mean ice concentration is

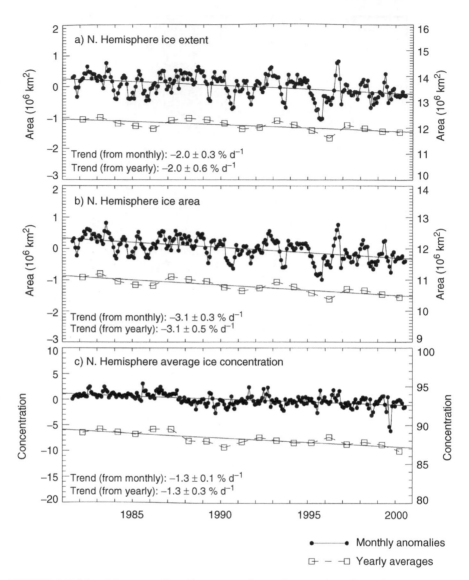

FIGURE 6.7 Monthly anomalies (departures from the mean) and yearly averages of (a) sea ice extent, (b) actual ice area, and (c) ice concentration, 1981–2000. (Reproduced from Comiso (2002). Reprinted from *Annals of Glaciology* with permission of the International Glaciological Society.)

decreasing at about 0.1% a^{-1} (Comiso 2002). Superimposed on these long-term trends appears to be a cyclic or quasiperiodic variation with a period of about 5 years (Comiso 2002) (Figure 6.7), perhaps related to the Atlantic Oscillation and El Niño Southern Oscillation phenomena (Deser, Walsh, and Timlin 2000; Dickson et al. 2000; Gloersen 1995). The perennial ice appears to be decreasing at twice the rate of the ice cover in general. There are significant regional variations: the largest reductions have occurred in the Arctic Ocean, the Kara and Barents Seas, and the Seas of Okhotsk and Japan, while increases (though probably not statistically significant) have occurred in the Bering Sea and the Gulf of St. Lawrence (Parkinson and Cavalieri 2002).

The first long-term analysis of trends in concentrations of multiyear and first-year ice was performed by Johannessen, Shalina, and Miles (1999), who showed a decrease of 0.7% a^{-1} in the area of multiyear ice in the Arctic Ocean from 1978 to 1998. This finding is supported by the observation, again using analysis of passive microwave data, of an increase of 8% (5 days) in the length of the melt season over roughly the same period (Smith 1998).

Comparison of ice extent and ice area trends with trends in surface temperature* has been carried out by Comiso (2002). Temperature trends for the period 1981 to 2000 are $+0.05 \pm 0.02$ K a^{-1} over sea ice areas, -0.01 ± 0.04 K a^{-1} over Greenland, and $+0.10 \pm 0.02$ K a^{-1} over other high-latitude land areas. In general, departures from the trends in sea ice and in surface temperature are negatively correlated.

Another way of specifying trends in sea ice extent is to consider the length of the sea ice season, defined, for example, as the number of days in a year for which the concentration exceeds 15%, 30%, or 50% (Parkinson 2002, 1994; Watkins and Simmonds 2000). The results of this type of analysis for the southern hemisphere show a lengthening of the ice season in the Ross Sea, the coast of East Antarctica and part of the Weddell Sea, and a shortening in the Bellingshausen and Amundsen seas and parts of the Weddell Sea. Overall, more areas show a positive than a negative trend (Parkinson 2002). It is clear that the Arctic and the Antarctic have not been varying synchronously over the last 20 years.

6.3 ICE TYPE

Ice type can be determined from VIR imagery if the spatial resolution is high enough to allow the texture to be resolved adequately, but in practice most approaches to spaceborne determination of ice type have focused principally on the use of microwave imagery. This is probably due to a combination of the susceptibility of VIR imagery to cloud cover and the fact that high-resolution VIR imagery is obtained from comparatively narrow swaths, which makes it unsuitable for synoptic studies of large areas. Nevertheless, VIR imagery continues to be useful for detailed investigation of sea ice, for example, the study of melt ponds (Fetterer and Untersteiner 1998).

The use of passive microwave imagery for determination of ice type has been described in the previous section, since the strong differences in emissivity

---

*Surface temperatures are estimated from satellite thermal infrared data (e.g., Steffen et al. 1993; Comiso 2000). See Section 6.7.

between first-year and multiyear ice mean that discrimination of ice type is a necessary step in the determination of ice concentration. The form of Figure 6.4 suggests that passive microwave algorithms are more sensitive to total ice concentration than to the relative proportions of first-year and multiyear ice, and we have already noted the sensitivity of the algorithms to the presence of melt features. Some of these ambiguities have been successfully resolved by incorporating *microwave scatterometer* (essentially a nonimaging radar) data (Voss, Heygster, and Ezraty 2003).

The analysis of synthetic aperture radar imagery has received considerable attention as a means of determining sea ice type. SAR offers the usual advantages (potentially high spatial resolution, independence of daylight, and largely independent of weather conditions), but interpretation is complicated by speckle, natural variability of the backscattering coefficient of sea ice (Beaven et al. 1997; Barber et al. 1995), especially as a result of metamorphic processes (Livingstone et al. 1987), and variability of the backscattering coefficient of open water as a function of roughening by wind (Kwok et al. 1992). Early attempts at classification were based on manual analysis and confined to the winter months (Kwok et al. 1992), although since the advent of the ERS satellites in the early 1990s, automated processes have been developed. SAR data collected at the Alaska SAR Facility (ASF) are routinely processed by the Geophysical Processor System (GPS) (Kwok, Cunningham, and Holt 1992). This was originally designed for ERS-1 data, which are smoothed to a resolution of 200 m to reduce speckle. The algorithm is purely radiometric, i.e., it takes no account of spatial context. The procedure is to cluster the image on the basis of the mean and standard deviation of the backscattering coefficient, then to label the most populous cluster on the basis of its closeness to values in a look-up table (LUT) (see Table 6.1 and Figure 6.8). Other ice types are then classified on the basis of the expected *differences* in mean backscatter (this allows for the effect of radiometric calibration errors). Ancillary meteorological data are used to make the selection from the LUT. As Table 6.1 implies, discrimination of open water from thin sea ice is difficult at C-band. Discrimination improves at higher frequencies (Matsuoka et al. 2002b).

Most spaceborne SAR imagery provides a single channel of data (single frequency and single polarization state), and this imposes a significant limitation on its use for classification. Various approaches have been adopted to try to circumvent this limitation. The ASF GPS algorithm in effect extracts two parameters from the data — the mean backscattering coefficient and the standard deviation, which can be regarded as a simple measure of image texture. Other approaches using the standard deviation or variance as a measure of image texture are due to Sun, Carlström, and Askne (1992); Kwok and Cunningham (1994). More sophisticated texture measures have also been adopted, for example using the gray-level co-occurrence matrix (Shokr 1991; Nystuen and Garcia 1992) and the autocorrelation function (Collins, Livingstone, and Raney 1997). In the latter case, a mathematically rigorous approach was adopted in which the spatial variability of the scene was

TABLE 6.1
Expected CVV Backscattering Coefficients at 25° Incidence Angle for Different Ice Types (Kwok, Cunningham, and Holt 1992)

Ice Type	Thickness (m)	σ^0 (dB)	Std Devn (dB)	A (dB/degree)[a]
Winter to Early Spring				
Multiyear	> 2.2	−8.6	2.2	−0.08
First-year	0.2–2.2	−14.0	2.1	−0.24
New ice/open water	< 0.2	< −18.0		
Late Spring				
Multiyear	> 2.2	−10.7	2.1	−0.27
First-year	0.7–2.2	−13.2	1.1	−0.22
New ice/open water	< 0.2	< −18.0		
Multiyear/first-year	> 0.2	> −16.0		
New ice/open water	< 0.2	< −18.0		
Summer				
Multiyear	> 1.5	−10.5	1.7	−0.04
First year	0.3–1.2	−12.5	1.9	−0.21
New ice/open water	< 0.3	< −18.0		

[a]Dependence of backscattering coefficient on incidence angle.

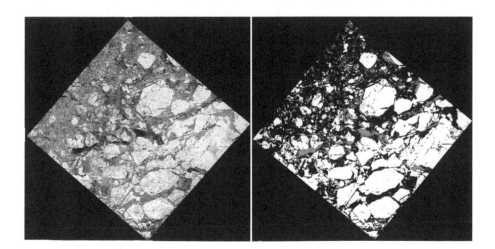

FIGURE 6.8 (Left) ERS-1 SAR image of part of the Beaufort Sea, 27 November 1991. (Right) Image classified as multiyear (white), first-year (black) and thin ice (gray). (Adapted from Kwok, Cunningham, and Holt (1992), Copyright American Geophysical Union 1992.)

separated from the system response. This is mathematically complicated but has better potential to derive unambiguous texture signatures. In an experimental investigation in the Labrador Sea marginal ice zone, Collins, Livingstone, and Raney (1997) concluded that single-channel SAR data (i.e., a single frequency and polarization state) is usually adequate to distinguish between different forms of a particular ice type (e.g., between smooth and rough new ice), although not between different ice types. The most generally useful single channel of SAR data is one of the copolarized X-band channels (HH or VV), though CHH is also valuable. For discrimination between different ice types, a combination of two radar channels, XHV and CHH or XHV and CHV, is required.

The utility of image texture can be enhanced if the image is first segmented into discrete homogeneous regions, representing individual floes (Skriver 1989; Soh and Tsatsoulis 1999). Bochert (1999) adopted this approach and used airborne scanner imagery (visible and thermal infrared) to train the radar classification and to assess its accuracy.

The analysis of multitemporal SAR imagery is useful during the melt period, since first-year ice and multiyear ice respond quite differently to the onset of melt (Thomas and Barber 1998): the backscattering coefficient of multiyear ice decreases (Winebrenner et al. 1994), while that of first-year ice increases (Livingstone et al. 1987) (Figure 6.9). Physical modeling of the ice is helpful in interpreting SAR data from the melt period (Beaven et al. 1997).

Other approaches to augmenting the data content of SAR imagery include merging it with other sources of imagery. Lythe, Hauser, and Wendler (1999) described a successful merging of SAR and AVHRR imagery from the Antarctic. The SAR data were first despeckled, then the image histogram was investigated to choose segmentation boundaries. Coregistered AVHRR thermal IR imagery was then used to provide a two-band (SAR + AVHRR4) classification, which improved the number of discriminable ice types (Figure 6.10).

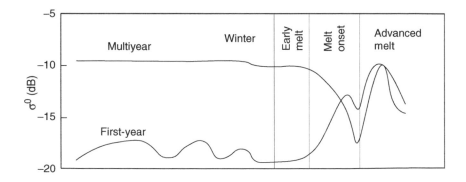

FIGURE 6.9 Schematic variation of the CVV backscattering coefficient of first-year and multiyear ice through the melt period. (Redrawn from Thomas and Barber 1998.)

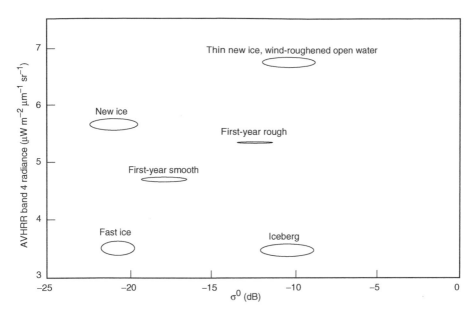

`FIGURE 6.10` Improved discrimination of ice types by combining ERS SAR data with AVHRR band 4 (thermal infrared) imagery. The ellipses represent one standard deviation in the data values. (Redrawn from Lythe, Hauser, and Wendler 1999.)

The most promising possibility for the determination of ice type from SAR imagery is the development of multifrequency and/or polarimetric SAR (Drinkwater et al. 1992). Figure 6.11 is adapted from Drinkwater et al. (1991). It shows an area of sea ice in the Beaufort Sea (11 March 1988) in C band (blue), L band (green) and P band (red). Labeled areas are multiyear floes (1 and 3), deformed first-year ice (2), and undeformed first-year ice (4 and 5). Spaceborne polarimetric SARs are now operational, although we may have to wait a few years before multifrequency systems are launched.

6.4 MELT PONDS AND SURFACE ALBEDO

During the summer, sea ice develops melt features such as ponds and wet-snow cover, which substantially lower the albedo (Grenfell and Maykut 1977). These can in principle be distinguished from multispectral optical imagery, since melt pond areas show relatively more reflectance in the blue part of the spectrum than wet-snow areas (Perovich, Maykut, and Grenfell 1986). Markus, Cavalieri, and Ivanoff (2002) demonstrate that the combination of spatial and spectral resolution provided by Landsat ETM+ is sufficient to enable areas of summer sea ice with high proportion of melt ponds to be distinguished. The onset of melt is also detectable in SAR imagery, as discussed in Section 6.3, and this can be used as an indirect method of estimating the albedo if VIR

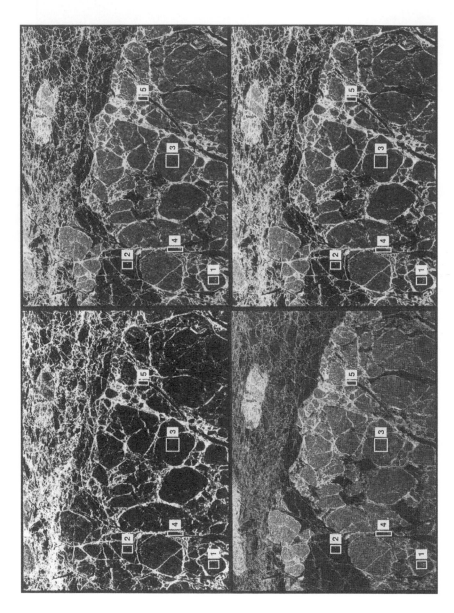

FIGURE 6.11 (See Color Figure 6.2 following page 108) Multifrequency airborne SAR image of sea ice in the Beaufort Sea. (Top left) P band; (top right) L band; (bottom left) C band; (bottom right) RGB composite of all three bands. (See text for details. Adapted from Drinkwater et al. (1992), Copyright American Geophysical Union 1992.)

imagery is not available (Thomas and Barber 1998; Barber and LeDrew 1994; Barber, Papakyriakou, and LeDrew 1994).

6.5   ICE THICKNESS

The thickness of sea ice is difficult to determine by remote sensing methods as we have defined them. Upward-looking sonar from submarines (Wadhams and Comiso 1992) or from fixed sensors, shipborne (or ship-based) observations, and measurements of elastic wave propagation in the ice pack can be used but these are all slow and limited in spatial coverage, and the results obtained from these methods are not currently consistent with one another. Estimates of the change in mean ice thickness of the Arctic Ocean between 1970 and 1990 currently range from −1.3 m (around −40%) (Rothrock, Yu, and Maykut 1999) to essentially zero (Winsor 2001; Nagurny, Korostelev, and Ivanov 1999).

Among the remote sensing techniques that can be used to measure ice thickness, airborne impulse radar is commonly used (Kovacs and Morey 1986), operating at a typical frequency of 100 MHz (Figure 6.12). However, impulse radar or ground-penetrating radar is only effective for ice thicknesses between about 1 m and 10 m. A related electromagnetic technique is the electromagnetic induction method (Holladay, Rossiter, and Kovacs 1990; Rossiter and Holladay 1994) in which transmitted radiation at 1 to 250 kHz induces eddy currents in the sea water below the ice. The reradiated field is attenuated by passage through the ice. The accuracy of this method is typically 0.1 m.

Airborne laser profiling was first applied to the problem of determining sea ice thickness in 1964 (Ketchum 1971), and an extensive survey was performed in 1987 (Comiso et al. 1991). Laser profiling measures the freeboard of the ice (the amount above the waterline), so a multiplicative factor must be applied to determine the ice thickness (Wadhams et al. 1992). An airborne study in the Sea of Okhotsk was described by Ishizu, Mizutani, and Itabe (1999) with an accuracy of better than 2 cm (Figure 6.13). Airborne laser profiling is, like all airborne techniques, suited primarily to small areas. Higher rates of areal coverage are promised by spaceborne laser profiling. Spaceborne radar altimetry has also begun to show promise as a technique for measuring ice thickness, through determination of freeboard, over large areas (Bobylev, Kondratyev, and Johannessen 2003).

Sea ice thickness can be estimated indirectly from SAR data. Matsuoka et al. (2002b) reported a significant correlation between the cross-polarized L-band backscattering coefficient and the ice thickness, fitted by the following regression formula (after first correcting the data for incidence angle variations)

$$\sigma^0_{LHV} = 7.3 \log\left(\frac{d}{m}\right) - 28.4 \text{ dB} \qquad (6.5)$$

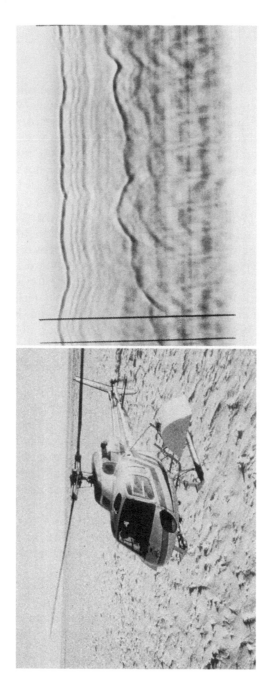

FIGURE 6.12 Helicopter-mounted impulse radar equipment (left) and output showing the ice top and bottom surfaces (right). (Reproduced from Wadhams and Comiso (1992), Copyright American Geophysical Union (1992), originally by courtesy of Dr. Gordon Oswald.)

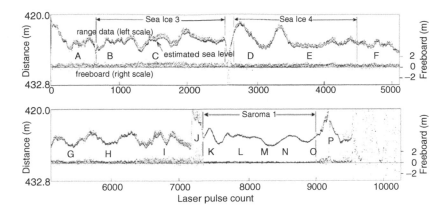

FIGURE 6.13 Airborne laser profiler data from sea ice in the Sea of Okhotsk and lake ice on Lake Saroma, Japan. The horizontal spacing of the pulses was 3.6 m. (Reproduced from Ishizu, Mizutani, and Itabe (1999) with permission of Taylor & Francis plc.)

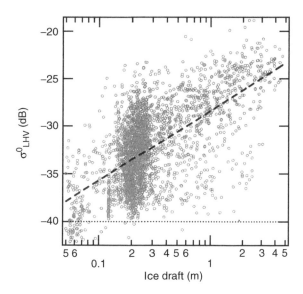

FIGURE 6.14 Correlation between L-band cross-polarized backscattering coefficient and ice draught. (Reproduced from Matsuoka et al. (2002b). Reprinted from *Annals of Glaciology* with permission of the International Glaciological Society.)

where d is the ice draught (Figure 6.14). The mechanism for this correlation, which is much less strong at higher frequencies (Wadhams and Comiso 1992), is probably the association between thickness and surface roughness. Melling (1998) found a strong correlation between the spatial frequency of ridges detected in SAR imagery of first-year ice and the mean ice draught. However, estimating the frequency of ridges in SAR imagery is complicated by the fact that young leads can also appear as filamentous features. The approach is less suitable for older ice because the ridges are harder to resolve.

Two other methods have been proposed for estimating ice thickness from SAR data. The first is to use the SAR data to classify the ice type and to use this as a proxy indicator of thickness (Melling 1998; Haverkamp, Soh, and Tsatsoulis 1995). This method relies on the fact that multiyear ice is significantly thicker — by a factor of about three — than first-year ice, and it can clearly be adapted to any technique capable of yielding reliable estimates of the extent of multiyear ice. The second method is to use a time-series of SAR images to determine the ice motion (see Section 6.6), and to model the growth of new ice in leads using Lebedev's formula:

$$h = 1.33F^{0.58} \tag{6.6}$$

where h is the ice thickness in cm and F is the cumulative number of degree–days of freezing (Maykut 1986). This is the procedure adopted by the Radarsat Geophysical Processor System (RGPS) (Kwok 2002).*

6.6 ICE DYNAMICS

Previous methods for determining the motion of sea ice have employed buoys drifting on the ice cover, as well as drifting ice-stations and ships (Liu and Cavalieri 1998). These approaches provide data of low spatial resolution, and coverage is particularly poor for the Southern Ocean. Satellite imagery is clearly an attractive alternative. However, the determination of ice motion requires sequential images separated by typically only a few days at most, and this means that narrow-swath sensors such as Landsat are generally unsuitable. Wide-swath VIR imagery such as AVHRR, and SAR imagery, is thus favorable for this application. All approaches using VIR imagery are, as usual, made difficult by cloud cover and by polar night, so the analysis of sequences of SAR images is now the preferred method (Johannessen et al. 1983; Campbell et al. 1987; Fily and Rothrock 1991). Passive microwave imagery generally has too coarse a spatial resolution to be of much use for determining ice motion, although the SSM/I channel at 85 GHz, with a resolution of 12.5 km, has been successfully analyzed by Liu and Cavalieri (1998) using wavelet transforms, which are like spatially and temporally localized Fourier transforms (Figure 6.15).

Early approaches to determining ice motion from satellite imagery relied on manual identification of features common to two or more images. A more

---

*http://www-radar.jpl.nasa.gov/rgps/radarsat.html.

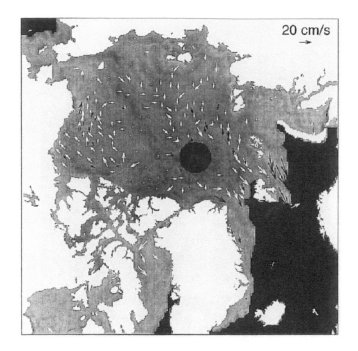

FIGURE 6.15 Sea ice motion vectors for 12 December 1992, determined by wavelet transform analysis of 85 GHz SSM/I data. The black arrows show displacements determined from buoys. (Reproduced from Liu and Cavalieri (1998) with permission of Taylor & Francis plc.)

recent approach is to compute the cross-correlation function between two images: the images are partitioned spatially, usually in a hierarchical way (starting with the largest-scale features), and the maximum cross-correlation coefficient is used to determine the relative motion of the ice represented by a region of the image (Holt, Rothrock, and Kwok 1992). This approach has been applied to AVHRR (Emery et al. 1991; Ninnis, Emery, and Collins 1986; Heacock et al. 1993) and SAR (Fily and Rothrock 1991; Collins and Emery 1988) imagery (Figure 6.16). It is most useful where ice concentrations are high and rotation of floes is therefore limited. In the marginal ice zone, where floes are freer to rotate, it is less successful.

Feature recognition algorithms have been applied to address the problem of rotation. The general principle is first to isolate a feature (a floe, lead, ridge, etc.) from the image by some kind of segmentation procedure, and then to characterize the feature by a number of parameters that do not depend explicitly on its orientation, so that the most likely corresponding feature can be recognized in a second image. This approach is complementary to the correlation technique, since it works better when the ice concentration is low and floes are well separated from one another. Leads are linear features that are comparatively easy to recognize automatically, and this fact was exploited

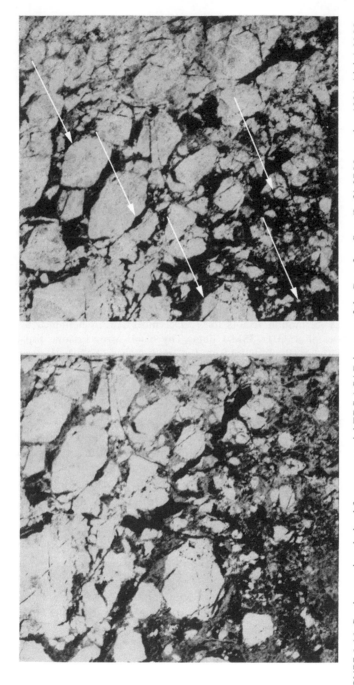

FIGURE 6.16 Sea ice motion derived from sequential ERS-1 SAR images of the Beaufort Sea. (Left) 27 November 1991; (right) 30 November 1991, with displacement vectors superimposed. (Modified from Holt, Rothrock, and Kwok (1992), Copyright American Geophysical Union 1992.)

by Vesecky et al. (1988). Ridges can also be used for tracking but are more difficult to recognize automatically (Vesecky, Smith, and Samadani 1990). The outline of a lead or floe can be represented by its ψ–s characteristic (Kwok et al. 1990). This is a graph that plots the variation of the orientation ψ of the tangent to the outline as a function of distance s measured along the edge from some arbitrary starting point (Figure 6.17). Apart from a shift in s, the graph is invariant under rotation of the feature. Similar approaches involve deriving orientation-independent shape parameters (e.g., the area, length of perimeter, and the semimajor and semiminor axes of the best-fitting ellipse).

The accuracy with which the absolute motion of an ice feature can be determined from sequential images depends on both the accuracy with which a position can be determined within each image and the relative registration accuracy between the two images. The latter can be improved if there are static features visible in both images that can be used as control points, but this situation does not usually arise in areas of sea ice cover except in the case of very wide swath imagery.

SAR interferometry has been demonstrated as useful for studying small motions of fast ice (Dammert, Leppäranta, and Askne 1998) (Figure 6.18).

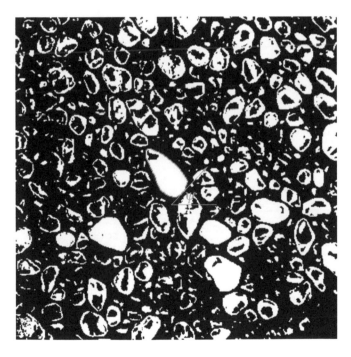

FIGURE 6.17 (Above) Thresholded image of sea ice (in fact, an early stage of ice development called *pancake ice*). The perimeter of the floe at the center has been outlined in gray. (Overleaf) The ψ (s) characteristic for the floe, showing how the angle ψ (in degrees) of the tangent to the perimeter varies with distance s (in pixels) measured along the perimeter.

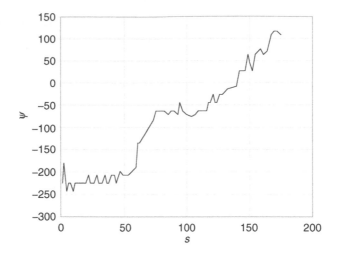

FIGURE 6.17 Continued.

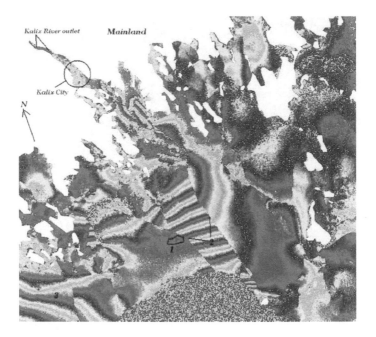

FIGURE 6.18 (See Color Figure 6.3 following page 108) Interferogram derived from ERS-1 SAR images of the northern end of the Gulf of Bothnia, 27 and 30 March 1992. One cycle of phase corresponds to a movement of 28 mm. The linear features at "2" are the tracks of icebreakers; the incoherent region at bottom center is open water. (Reproduced from Dammert, Leppäranta, and Askne (1998) with permission of Taylor & Francis plc.)

This suits it to studying ice mechanics during the initial transition from static to moving ice.

6.7   ICE TEMPERATURE

The surface temperature of sea ice can be determined using thermal IR sensors, such as AVHRR, but cloud cover is a serious problem. Passive microwave data are also valuable. The NASA Team algorithm exploits the fact that the emissivities of first-year and multiyear ice are essentially identical at 6.5 GHz (Comiso 1983) to eliminate errors arising from uncertainty about the ice type. The formula for the ice temperature T_i then becomes

$$T_i = \frac{T_{6.5V} - (1 - C)T_{w,6.5V}}{C\varepsilon} \tag{6.7}$$

where $T_{6.5V}$ is the observed brightness temperature in 6.5 GHz vertically polarized radiation, $T_{w,6.5V}$ is the corresponding value for open water, C is the total ice concentration, and ε is the emissivity. T_i is set to 271 K whenever C is less than 0.8 on the grounds that ice near open water is likely to be near the melting temperature. Other algorithms are discussed by Steffen et al. (1992).

7 Remote Sensing of Freshwater Ice

7.1 INTRODUCTION

The problems of remote sensing of freshwater (i.e., lake and river) ice are broadly similar to those for sea ice, although freshwater ice differs physically from sea ice by its lack of salt content. However, a major difference is the spatial scale of the phenomena. While seas and oceans are hundreds or thousands of kilometers in extent, the width of most rivers and lakes is of the order of kilometers or less. This has caused remote sensing of the ice cover of rivers and small lakes to be hampered until comparatively recently by difficulties in obtaining adequate spatial resolution.

7.2 EXTENT

In general, visible–near-infrared imagery is well suited to the observation of freshwater ice (see Color Figure 7.1 following page 108) (Hall 1993; Borodulin and Prokacheva 1983; Dobrowolski and Gronet 1990; Michel 1971; Gatto and Daly 1986; Campbell et al. 1975; Foster, Schultz, and Dallam 1978; Starosolszky and Mayer 1988). Figure 7.1 shows the reflective bands (1 to 5 and 7) of a Landsat TM image of an Arctic lake with partial ice cover. The contrast between ice and open water is strong in bands 1 to 4, though practically nonexistent in bands 5 and 7. The contrast between ice and vegetated terrain is strong in the visible bands (1 to 3), while the contrast between ice and snow-covered terrain (at the top left of the image) is strongest in the near infrared (band 4). Of the spectral bands provided by the earlier Landsat MSS sensors, the red band proved the most useful for identifying variations in river ice (Gatto 1990). Naturally, the usual restrictions on the availability of usable satellite data to periods of daylight and cloud-free conditions apply, and the temporal resolution imposed by a satellite's repeat period may also restrict the availability of data. The only ice type that is difficult to detect against a background of open water is "black ice."

Rivers and smaller lakes can be monitored using Landsat TM or ETM+ imagery, or something with similar spatial resolution. The coarser resolutions of AVHRR or MODIS are suitable for large lakes and provide better temporal resolution. AVHRR imagery was shown to be suitable for monitoring ice cover on Lake Baikal (Semovski, Mogilev, and Sherstyankin 2000). The thermal

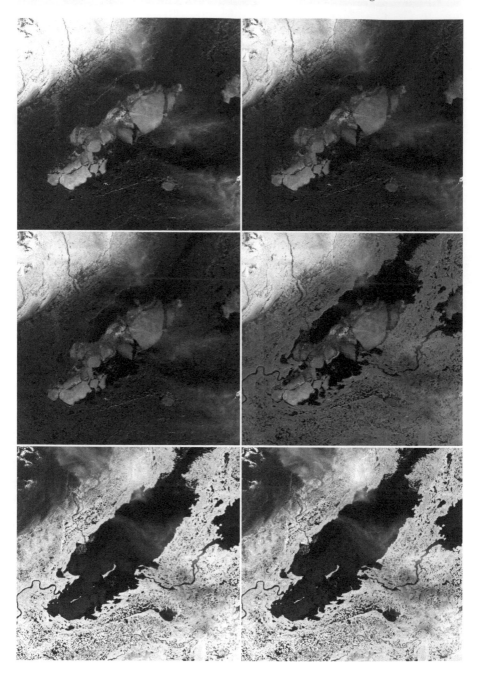

FIGURE 7.1 Extract of a Landsat 5 TM image centered on Lake Melkoye (69° 20′ N, 89° 15′ E) in the Russian Arctic. The image was acquired on 9 July 1995, and the extract covers an area approximately 40 km square. (Top row) bands 1 and 2; (middle row) bands 3 and 4; (bottom row) bands 5 and 7.

FIGURE 7.2 Thermal infrared channel (band 6) of the image shown in Figure 7.1. The area of ice cover can be discriminated from open water through an abrupt change in brightness temperature.

infrared channels of these instruments have also been exploited (Figure 7.2 shows the contrast between ice and open water in the thermal infrared band of a Landsat TM image). Dates of freeze-up and break-up of the ice on large (greater than 1000 ha) lakes in Wisconsin were estimated by Wynne and Lillesand (1993) using AVHRR data and a criterion of temperatures below −2°C to define the presence of ice. Comparison with *in situ* measurements showed that while the date of break-up was estimated to within a few days without systematic bias, the date of freeze-up tended to be placed about 5 days later than the true date.

Imaging radar has been used to study freshwater ice since the early 1970s (Borodulin 1989; Melloh and Gatto 1990; Sellmann, Weeks, and Campbell 1975; Weeks, Sellmann, and Campbell 1977). The formation of river ice can usually be detected through an increase in the backscattering coefficient. This is illustrated by Figure 7.3, which shows a SAR image of the Burntwood River, Manitoba, in January 1990. Shore-fast ice exhibits a high backscattering coefficient, while open water, including the central part of the river, which

FIGURE 7.3 (Above) Airborne CHH SAR image of Burntwood River, Manitoba, January 1990. (Below) Oblique aerial photograph of the First Rapids area, showing shore-fast ice and a region of frazil slush flowing down the middle of the river. (Reproduced from Leconte and Klassen (1991), with permission of the Arctic Institute of North America.)

contains flowing slush, exhibits a lower backscattering coefficient. Figures 7.4 and 7.5 show SAR images of Split Lake, Manitoba, in April 1989. The backscattering coefficient at the western end of the lake is high, while that in the central part is low. The low backscatter is attributable to a smooth ice surface, while the higher backscatter at the western end is due to congealed

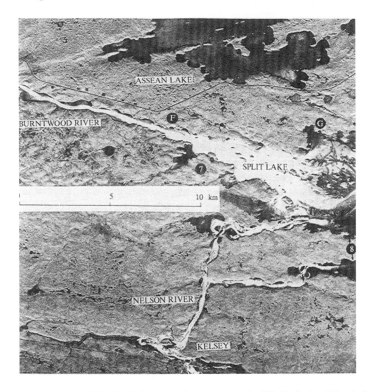

FIGURE 7.4 Airborne CHH SAR image of western end of Split Lake, Manitoba, April 1999. (Reproduced from Leconte and Klassen (1991), with permission of the Arctic Institute of North America.)

frazil slush that has been transported into the lake from the Burntwood River (Leconte and Klassen 1991).

The response of SAR to lake ice is strongly affected by the depth of the lake (Hall et al. 1994; Saich, Rees, and Borgeaud 2001). Typical shallow lake ice consists of clear ice on top (a few tens of centimeters), underlain by ice containing tubular bubbles (Morris, Jeffries, and Weeks 1995; Jeffries et al. 1994). Observations of CHH backscatter from shallow Alaskan lakes show an initially low backscatter (typically $-20\,dB$), which generally increases as the winter progresses, such that by late November (when it reaches typically $-10\,dB$) the contrast with the surrounding terrain has been lost (Figure 7.6). Brighter linear features, probably related to ice deformation, are also visible. Later in the winter (up to late January), lakes often show a marked decrease (around $10\,dB$) in backscatter, possibly caused by the ice freezing to the bottom. SAR data have proved useful for determining whether a lake or river is frozen throughout (Chacho, Arcone, and Delaney 1992; Leconte and Klassen 1991; Wakabayashi, Jeffries, and Weeks 1992). The most important mechanism appears to be the strong dielectric contrast between ice and water (3.2 cf. 80) compared with the much weaker contrast between ice and (frozen)

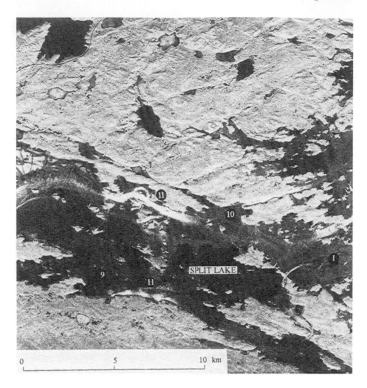

FIGURE 7.5 Airborne CHH SAR image of central Split Lake, Manitoba, April 1989. (Reproduced from Leconte and Klassen (1991), with permission of the Arctic Institute of North America.)

sediment ($\varepsilon \approx 8$) (Elachi, Bryan, and Weeks 1976), although the presence of scattering air bubbles also contributes (Weeks et al. 1978). The behavior of the backscatter during the spring thaw appears to be more complicated (Morris, Jeffries, and Weeks 1995).

Observations of CHH SAR data from deep lakes generally exhibit the same behavior as shallow lakes (backscatter increasing from a low value at the start of the winter) although without the subsequent decrease associated with freezing to the bottom. Deformation features can be more prominent (e.g., Figure 7.7) and zones of high-backscatter material can appear, probably deformed ice brought in from rivers (e.g., Figure 7.4).

7.3   CLASSIFICATION OF ICE TYPE

VIR imagery such as Landsat can be used to classify ice types. Early studies using Landsat 1 data (Leshkevich 1981, 1985) allowed the following lake ice types to be distinguished: new ice, consolidated pack, slush, brash, flow, snow-covered ice, and wet-snow-covered ice, as well as open water. On rivers, aufeis

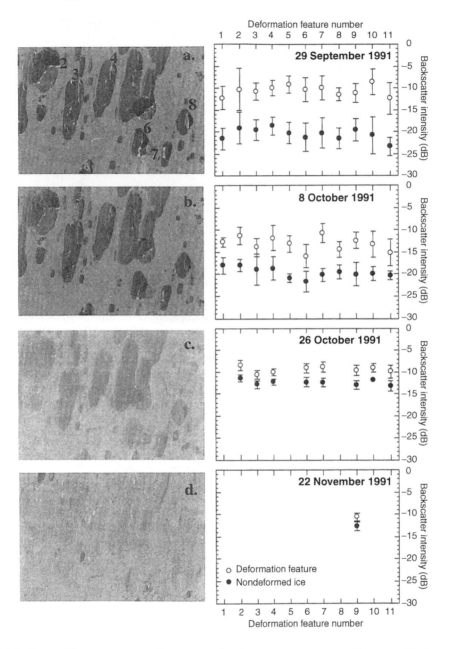

FIGURE 7.6 Time-sequence of ERS-1 SAR image extracts near Barrow, Alaska, showing the variation in backscatter of the numbered deformation features and the undeformed lake ice. The image extract covers an area approximately $21 \times 15\,\text{km}$. (Reproduced with permission from Morris, Jeffries, and Weeks (1995); SAR data copyright ESA 1991.)

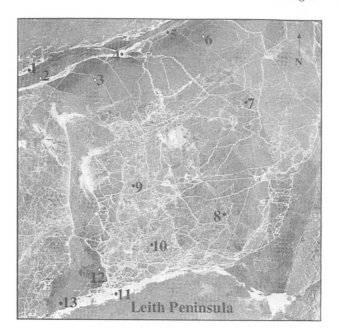

FIGURE 7.7 Deformation features in Great Bear Lake, Alaska, shown in ERS-1 SAR imagery. The extract covers an area approximately 70 km square. (Reproduced with permission from Morris, Jeffries, and Weeks (1995); SAR data copyright ESA 1991.)

can also be distinguished (Hall and Roswell 1981). For very large bodies of fresh water such as the Great Lakes, AVHRR imagery has proved effective (Wiesnet 1979).

SAR imagery can be used to classify ice in large water bodies. In general, smooth ice gives low returns, rough ice gives higher returns, and this, coupled with the presence or absence of deformation features, forms the basis of classification. The ability of SAR to identify a number of different types of lake ice has been recognized since the 1970s (Bryan and Larson 1975; Larrowe et al. 1971; Parashar, Roche, and Worsfold 1978). A similar classification of river ice can be performed (Gatto 1993). Lower frequency (L-band) SAR appears to give better results than higher frequencies (Melloh and Gatto 1990).

Passive microwave radiometry has also been demonstrated as effective for ice monitoring (Swift, Harrington, and Thornton 1980), although this is unlikely to be a useful technique from spaceborne observations except for the largest of water bodies because of the low spatial resolution.

7.4   ICE THICKNESS

The determination of freshwater ice thickness is attended by many of the same problems as for sea ice. Surface or airborne impulse radar measurements

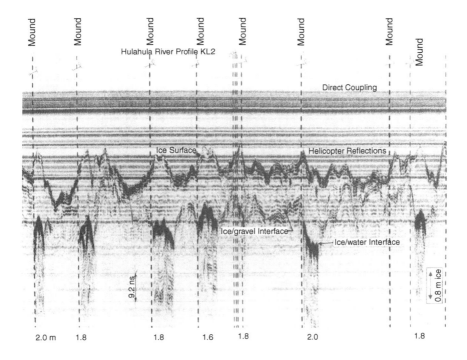

FIGURE 7.8 Output from helicopter-borne impulse radar survey of the Hulahula River, Alaska, in March 1989, showing returns from the ice surface, and ice/bottom and ice/water interfaces. (Reproduced from Arcone, Delaney, and Calkins (1989) with permission of the Cold Regions Research and Engineering Laboratory.)

(Annan and Davis 1977; Arcone and Delaney 1987) are effective,* although the large beamwidth of such instruments (70° is typical) means that they must be kept close to the surface. At a frequency of 600 MHz a thickness resolution of 0.2 m is typical (Arcone, Delaney, and Calkins 1989) (Figure 7.8). Interference of the radiation with the metal structures of the aircraft can be a problem. A somewhat similar approach to what is essentially *in situ* measurement is the use of continuous-wave millimeter-wave (26.5 to 40 GHz) radar (Yankielun, Arcone, and Crane 1992; Yankielun, Ferrick, and Weyrick 1993). This is usually helicopter-borne, with the sensor suspended very close (3 to 10 m) to the surface. Ice thickness can be also determined by airborne laser profiling (e.g., Krabill, Swift, and Tucker 1990; Ishizu, Mizutani, and Itabe 1999). The first such measurement was in 1964 (Ketchum 1971). As with sea ice, the idea is to measure the freeboard, since the probability distribution function for the

---

*Impulse radar measurements are also useful to determine whether river or lake ice is frozen to the bed (Arcone, Delaney, and Calkins 1989), as shown in Figure 7.8.

thickness is essentially just a scaled version of that for the freeboard (Wadhams et al. 1992).

Passive microwave radiometry can reveal information about ice thickness (Schmugge et al. 1974). At lower frequencies, the attenuation length is greater so the observed brightness temperature is more strongly influenced by the underlying water. High correlation is observed between ice thickness and brightness temperature at 5 GHz (Hall et al. 1981). As usual, though, the spatial resolution of passive microwave radiometry makes it unfit for use as a spaceborne technique.

Indirect approaches to the estimation of ice thickness have also been adopted, for example, through the relationship between thickness and roughness or by relating the length of time for which an ice cover is retained into the summer to the maximum ice depth (Sellmann et al. 1975).

7.5   ICE MOTION

The principles for determining motion of freshwater ice are similar to those for sea ice and are not discussed here in detail. In the case of river ice, the higher flow speeds relative to those encountered in lakes and seas mean that shorter time intervals are needed between successive images for feature-tracking to be successful, and this makes spaceborne observations unfit in most cases. However, aerial photography can be used to determine floe velocities. This was demonstrated by Starosolszky and Mayer (1988) who used two aerial photographs of the river Danube taken 7 minutes apart in February 1985.

8 Remote Sensing of Glaciers, Ice Sheets, and Ice Shelves

8.1 INTRODUCTION

The properties of a terrestrial ice mass that can be measured using remote sensing methods include its spatial extent, surface topography, bottom topography, total volume (which can be deduced from the surface and bottom topographies), surface flow field, accumulation and ablation rates (and hence mass balance), surface zonation, albedo, and changes in these quantities over time. As well as its importance in indicating the total amount of ice, surface topography provides important clues about the internal structure and can reveal flow features and grounding lines (Bamber and Bentley 1994).

The sensitivity of glaciers to climate, discussed in Chapter 1, has meant that a major focus of research has been the assessment of mass balance. Various approaches have been adopted, including direct measurement, indirect approaches, for example based on estimating the altitude of the equilibrium line, and modeling. Modeling based on energy balance calculations requires accurate measurements of the surface albedo (van de Wal, Oerlemans, and van der Hage 1992). However, since the principles of albedo measurement were discussed in Chapter 5 they are not repeated in this chapter.

Visible–near-infrared (VIR) and synthetic aperture radar (SAR) imagery both play major roles in the remote sensing of terrestrial ice masses. VIR imagery benefits from the high albedo of snow, making it particularly easy to recognize, and the intuitive interpretability of images. On the other hand, the imagery is confined to daylight, cloud-free conditions, which can prove a major limitation. Radar imagery does not suffer from these constraints, but is subject to a number of complications, such as geometric distortion and speckle, which were discussed in Chapter 2. The interaction mechanisms between microwave radiation and a glacier are also complicated and varied, which inhibits intuitive interpretation and introduces some ambiguity.

Other remote sensing techniques are also important. Radio echo-sounding and closely related methods such as ground-penetrating radar can reveal depth and internal structure, and SAR interferometry reveals surface topography and velocity. Recently, laser profiling has demonstrated the ability to study topography in great detail, fine enough to resolve subtle surface features.

Remote sensing of terrestrial ice masses (glaciers, ice caps, and ice sheets) is generally a well-developed field. Remotely sensed data are regularly used to compile inventories and surveys of glaciers in Europe and the United States. The highest information density and most complete historical record for mountain glaciers are in the European mountain ranges, especially the Alps (Kääb et al. 2002). The situation in South America is much less complete (Warren and Sugden 1993). The entire coastal zone of Antarctica from the 1970s onward is being mapped (Ferrigno et al. 1998; Williams et al. 1995). Since 1999, the international GLIMS* (Global Land Ice Measurements from Space) project has been carrying out routine monitoring of glaciers worldwide, using data from the ASTER and ETM+ sensors (Kargel 2000).

8.2   SPATIAL EXTENT AND SURFACE FEATURES

The study of glaciers by remote sensing methods dates back at least to the 1930s when aerial photography was applied to the problem (Williams and Hall 1993), and aerial photography still has considerable value. For example, the glacier inventory for Switzerland was compiled for 1973 using aerial photography (Müller, Caflisch, and Müller 1976). In keeping with the general approach of this book, however, we try to focus as much as possible on spaceborne techniques. The natural extension of aerial photography to the spaceborne domain is the use of VIR imagery, and this has proved exceptionally valuable for the study of terrestrial ice masses. The choice of swath width and spatial resolution is governed by the spatial scale of the phenomena to be investigated, with wide-swath instruments, such as AVHRR, providing a good match to the requirements of studying the large ice sheets of Antarctica and Greenland. Figure 8.1 shows a mosaic of AVHRR images covering the whole of Antarctica (Merson 1989). This mosaic revealed the complexity of the Antarctic ice sheet surface and provided some qualitative topographic information through shape-from-shading (discussed in detail in Section 8.3) at scales to about 10 km (Bindschadler 1998). Its ability to reveal the presence of surface features indicative of ice dynamics was demonstrated by Bindschadler and Vornberger (1990) and Casassa and Turner (1991). Discrimination between snow and cloud was a problem in the earlier AVHRR imagery (which lacked the 1.6 μm channel — see Section 5.2.1). The edge of the ice sheet can also be hard to discern, where it is formed by the boundary between land ice and fast sea ice. Higher-resolution data are valuable in this regard. The nominal resolution of AVHRR data is only 1 to 2.5 km. However, the historical archive of data is so large that some authors (Albertz and Zelianeos 1990) have

---

*http://www.glims.org/.

FIGURE 8.1 AVHRR band 1 mosaic of Antarctica.*

reported success in enhancing the resolution by about a factor of 2 by using several images with slightly different geometric registrations.

Higher-resolution satellite data, such as Landsat or ASTER imagery, and declassified satellite reconnaissance photographs provide historical data runs extending back to 1972 and about 1960, respectively, and satellite image atlases of Antarctica and Greenland have been compiled using these higher-resolution datasets (Swithinbank 1988; Weidick 1995). The roughly 100-fold improvement in spatial resolution naturally shows very much more detail, including crevasses and grounding lines (the transition between grounded and floating ice) (Jacobel, Robinson, and Bindschadler 1994), surface melt features (Vornberger and Bindschadler 1992), supraglacial lakes (Wessels, Kargel, and Kieffer 2002), sastrugi, snow dunes, pitted patterns, glazed surfaces (Watanabe 1978; Goodwin 1990) and "snow megadunes," occupying more than 500,000 km² and oriented perpendicular to the regional direction of the katabatic winds (Frezzotti et al. 2002; Fahnestock et al. 2000). As with the coarser-resolution imagery, surface features can be identified through a photoclinometric approach in satellite photography or high-resolution VIR imagery.

---

*http://terraweb.wr.usgs.gov/TRS/projects/Antarctica/AVHRR/.

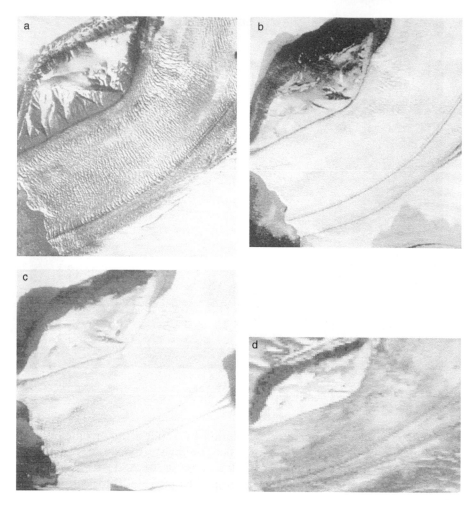

FIGURE 8.2 Kronebreen and Kongsvegen glaciers, Svalbard, images by (a) KFA-1000, (b) SPOT HRV, (c) Landsat TM, (d) Landsat MSS. (Reprinted from Dowdeswell et al. (1993), with permission of the International Glaciological Society.)

This has been demonstrated by, e.g., Dowdeswell et al. (1993) using Russian KFA-1000 photographs* (see Figure 8.2), which have an intrinsic spatial resolution of about 2 m (Baxter 1991). However, we should recall the compromise between spatial resolution and spatial coverage in VIR imagery. A single Landsat image covers an area of approximately 30,000 km², and it has been noted that it would require about 500 cloud-free images to cover the whole of Antarctica to 82.5° S (Thomas 1993).

---

*However, data from this particular instrument are noisy and uncalibrated so that quantitative analysis is difficult.

It was noted above that the Swiss glacier survey for 1973 was compiled using aerial photography. The equivalent inventory for 2000 is being compiled using satellite imagery (Paul et al. 2002). This offers the advantages of reductions in costs and manpower, plus the ability to monitor the small glaciers ($< 1 \, \text{km}^2$) that represent about 24% of the total glacier area in Switzerland (Kääb et al. 2002). Techniques for the delineation of glaciers using Landsat imagery include the following:

1. Forming a ratio image between two TM bands and segmenting it (e.g., Bayr, Hall, and Kovalick 1994; Hall, Chang, and Siddalingaiah 1988; Jacobs, Simms, and Simms 1997; Rott 1994)
2. Unsupervised classification (e.g., Aniya et al. 1996)
3. Supervised classification (e.g., Gratton, Howarth, and Marceau 1990; Sidjak and Wheate 1999; Casassa et al. 2002)

Paul et al. (2002) concluded that the most accurate results were obtained by thresholding a ratio image of Landsat TM5/TM4 DN data (i.e., not corrected to spectral radiance) (see Figure 8.3). Other methods were too much affected by shadow, although all methods failed to discriminate debris-covered ice because of its spectral similarity to surrounding terrain. Median filtering somewhat improved the accuracy, but at the expense of limiting the smallest reliably detectable glacier to 10 hectares.

The ability of VIR imagery to provide a clear indication of the extent and surface features of a glacier also suits it for change-detection through the analysis of time-series of images. This technique has been widely applied, for example to the Barnes ice cap (Jacobs, Simms, and Simms 1997), glaciers in Austria and Iceland (Hall, Williams, and Bayr 1992), the Jakobshavn glacier (Sohn, Jezek, and Van der Veen 1998), the Sør Rondane mountains (Pattyn and Decleir 1993), the Larsen ice shelf (Skvarca 1994), James Ross Island (Skvarca, Rott, and Nagler 1995), and the Mertz and Ninnis Glaciers (Wendler, Ahlnäs, and Lingle 1996).

Imaging radar is also effective for determining the spatial extent of terrestrial ice masses. As was done with AVHRR imagery in the 1980s, a mosaic of SAR imagery was constructed for Antarctica in the 1990s using Radarsat data from the Antarctic Mapping Mission (Choi 1999) (Figure 8.4). C-band imagery can be used to map wet snow and ice-free surfaces but provides poor discrimination between glacier ice, snow, and bare rock. The discrimination of ice from bare rock can be improved by using a texture parameter, as demonstrated by Sohn and Jezek (1999), who used a simple coefficient of variance. L-band imagery distinguishes snow or ice from other surfaces (Shi and Dozier 1993). The discrimination is improved if the data are corrected for incidence-angle variations. A common model is to assume that the backscatter varies with incidence angle as

$$\sigma^0(\theta) = \sigma^0(0)(\cos \theta)^n \tag{8.1}$$

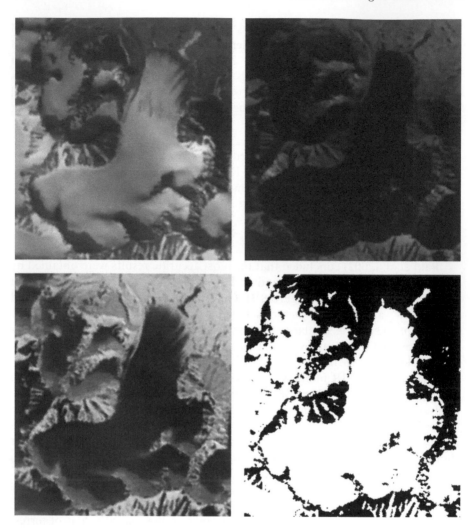

FIGURE 8.3 (Top left) Band 4 Landsat ETM+ image of Midre Lovénbreen, Svalbard; (top right) corresponding band 5 image; (bottom left) ratio of band 5 to band 4; (bottom right) thresholded ratio, showing approximately the extent of the glacier.

(Ulaby, Moore, and Fung 1982), where n is 1 for a perfectly rough surface (this is the Muhleman model of equation (5.24)), 2 for a volume-scattering material, and typically around 1.5 for a glacier surface (Shi and Dozier 1993).

The utility of monitoring variations in the margin of an ice sheet as a proxy for its dynamics has been emphasized by Sohn and Jezek (1999), who have developed an automated technique that combines imaging radar and high-resolution optical data. When this method was applied to ERS-1 SAR and SPOT imagery of the margin of the Jakobshavn Glacier in Greenland covering the period 1988 to 1992, the authors showed that the margin undergoes an

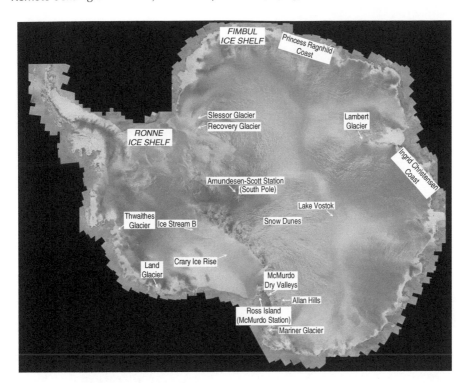

FIGURE 8.4 AMM (Antarctic Mapping Mission) Radarsat mosaic of Antarctica.* The original mosaic was constructed at a resolution of 25 m.

annual fluctuation of about 12 m. The technique has an absolute locational accuracy of about 200 m. Fricker et al. (2002) have used Radarsat imagery to monitor the position of the front of the Amery ice shelf between 1997 and 2000, and combined this with estimates from the 1980s and 1970s using Landsat data, and earlier dates using airborne and shipborne observations, to estimate the flow rate (about 1400 m/yr) and hence iceberg calving cycle (about 60–70 years) of the ice shelf. Another technique is *slant-range analysis* of radar altimeter data (Martin et al. 1983; Zwally and Brenner 2001; Zwally et al. 2002). This is based on the fact that, when a spaceborne radar altimeter encounters an abrupt change in surface elevation, such as that between open or ice-covered ocean and an ice shelf, it will continue to track the earlier surface for about a second after it has crossed the boundary. This results in a characteristic time-variation of the measured range, which can be used to determine the location of the margin. Like other methods based on measuring differences in the location of the boundary over time, it tends to underestimate the rate of change of area because it fails to detect small surge or calving events (Keys, Jacobs, and Brigham 1998). On the other hand, the technique is well

*http://www.space.gc.ca/asc/eng/csa_sectors/earth/radarsat1/antarctic.asp.

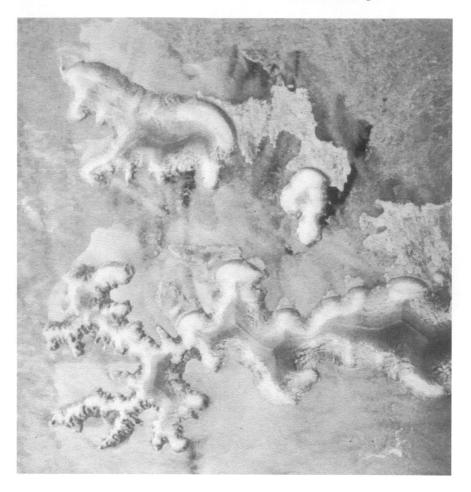

FIGURE 8.5 ERS-1 SAR image of Alex Land and George Land, Franz Josef Land, Russia, 04.09.1991, showing ice divides. (Copyright ESA 1991.)

suited to monitoring of a very large proportion of the margin of the Antarctic ice sheet (Zwally et al. 2002).

Like VIR imagery, SAR imagery also reveals surface features, including surface melt features (Vornberger and Bindschadler 1992), and features characteristic of surge-type glaciers (Dowdeswell and Williams 1997), ice divides, and drainage basins (Dowdeswell, Glazovsky, and Macheret 1995) (Figure 8.5).

8.3   SURFACE TOPOGRAPHY

The longest established remote sensing technique for measuring the surface topography of a glacier is *stereophotography*. As discussed in Section 2.2.4, this is capable of achieving a height accuracy of the order of $H/1000$, where H

is the height above the surface from which the photographs are obtained. Typical minimum flying heights are of the order of 1000 m, so a representative accuracy is 1 m. An example is given by Lundstrom, McCafferty, and Coe (1993) using vertical aerial photographs at a scale of 1:12,000 to give an estimated height accuracy of 0.5 m. By repeating the stereophotography after a few years, changes in glacier thickness can also be determined. The principal disadvantages of stereophotography are the limited spatial coverage (typically, a stereopair providing a height accuracy of 1 m will cover an area of the order of 1 km^2) and the difficulty of identifying conjugate points (common features) in the two photographs.

Although some airborne and spaceborne VIR sensors offer a stereo viewing capability, the achievable height accuracies are generally poorer than can be obtained from photographs taken from the same altitude. An alternative approach using VIR imagery is through *photoclinometry*, also referred to as shape-from-shading (e.g., Rees and Dowdeswell 1988; Wildey 1975; Bindschadler and Vornberger 1994; Bingham and Rees 1999; Bindschadler et al. 2002). This technique is based on the assumption that the bidirectional reflectance distribution function (BRDF) of the surface is uniform so that variations in the radiance detected by the sensor are due solely to variations in the viewing geometry. Since the illumination geometry (from the Sun) is constant, these variations are in turn due to variations in the surface slope. Qualitatively, the principle is straightforward: surfaces that face toward the Sun will be brighter than those that face away from it. The effect is enhanced by low solar elevations, but on the other hand, this increases the probability that some parts of the surface will be in shadow. For optical photoclinometry, the governing equation is (Bindschadler et al. 2002)

$$D = C \int T(IR \cos\theta - L_0 + S)d\lambda \qquad (8.2)$$

where D is the image brightness in sensor units, T is the bandpass transmittance, I is the exoatmospheric solar radiance, R is the surface reflectance, θ is the angle between the surface normal and the solar direction, L_0 is the radiance when $D=0$, and S is the path radiance. C is the calibration coefficient of the sensor. T, I, L_0, and S are functions of the wavelength λ. The integration is carried out across the sensor's bandpass. More simply, the relationship can be approximated as

$$D = A \cos\theta + B \qquad (8.3)$$

where A and B are constants that can be derived from ancillary data, such as the heights of a few known points, and it has been assumed that the surface reflectance R is constant.[*] In principle, either of the equations can be used

---

[*]In this case, it is important to include only areas in direct sunlight and for which the reflectance is uniform.

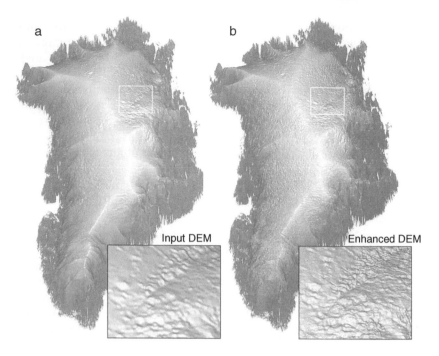

FIGURE 8.6 Visualizations of (a) original DEM of Greenland and (b) DEM after the inclusion of more detailed topography derived from photoclinometry. (Reprinted from Scambos and Haran (2002), with permission of the International Glaciological Society; original DEM from Bamber, Layberry, and Gogineni (2001)).

to determine θ from D and hence the surface slope, which can be integrated to generate the surface profile.

Most applications of photoclinometry use high-resolution imagery. However, Scambos and Fahnestock (1998) and Scambos and Haran (2002) have demonstrated the use of photoclinometry from AVHRR imagery to enhance the resolution of a DEM derived from radar altimetry (Figure 8.6). AVHRR channel 1 data are preferred owing to the lower sensitivity to grain size (Dozier, Schneider, and McGinnis 1981).

Radar photoclimometry is also possible, although the conditions to ensure a slope-dependent backscatter unaffected by other parameters are more restrictive, essentially requiring wet-snow conditions without significant melting (e.g., Vornberger and Bindschadler 1992).

The earliest spaceborne topographic technique is *radar altimetry*. The first satellite to give data useful for the large ice sheets was Seasat (1978), followed by Geosat (1985 to 1992), and the ERS and ENVISAT missions (1991 onward). The Cryosat mission (scheduled for launch in 2005) will be dedicated to radar altimetry over ice. Since the launch of ERS-1 in 1991, high-latitude radar altimeter coverage has been essentially continuous. The effective spatial resolution of the technique is typically about 1 km (Wingham 1995), and height resolutions approaching 1 cm are now possible, though 10 cm is a more usual figure. The normal method of removing systematic biases in the data is to use *crossover analysis*, in which the crossing points of spatially adjacent ascending and descending orbits are used to constrain the data (Zwally et al. 1989). Radar altimetry has been used to construct DEMs for large ice masses (e.g., the Antarctic DEM compiled from ERS-1 data (Liu, Jezek, and Li 1999)) although its spatial resolution makes it unfit for the study of small glaciers. The precision of topographic data compiled from radar altimetry is great enough to reveal the presence of some subglacial features, such as bedrock configuration (Section 8.4) and subglacial lakes, such as Lake Vostok (Kapitsa et al. 1996). Careful analysis of time series of altimetric datasets allows long-term changes in the ice thickness to be investigated. For example, Wingham et al. (1998) analyzed data from the ERS-1 and ERS-2 radar altimeters over Antarctica between 1992 and 1996. They observed significant spatial variation, and a mean rate of change of elevation of -9 ± 5 mm/year. A similar investigation by Khvorostovsky, Bobylev, and Johannessen (2003) for the Greenland ice sheet found a mean rate of -22 mm/yr for the period 1992 to 1996, using data from ERS-1, and a rate of $+108$ mm/yr for the period 1995 to 1999, using data from ERS-2. Combining the data from both satellites, and correcting for biases in the two instruments, the authors derived a mean rate of $+43$ mm/yr for the period 1992 to 1999. This is not spatially uniform, since the steeper marginal regions were found to be decreasing in elevation, while the interior is thickening, especially in the south-western part.

Because surface slopes on glaciers and ice sheets can easily reach a few degrees, slope correction, as discussed in Section 2.7.3, is often necessary (Brenner et al. 1983; Bamber 1994). Over particularly steep terrain (e.g., rugged unglaciated terrain adjacent to an ice sheet or glacier) the altimeter can suffer from "loss of lock" in which the instrument fails to predict the time at which the return signal will be received. In this case it must go into a reacquisition procedure, during which no useful data are retrieved. This is particularly undesirable for topographic measurements of ice sheets because it means that data are less likely to be acquired from the steeper marginal regions which are most sensitive to the effects of climate change (Khvorostovsky, Bobylev, and Johannessen 2003). Spaceborne radar altimeters from ERS-1 onward have been designed to be less sensitive to this phenomenon, at the expense of reduced height resolution. Over horizontal surfaces, elevations derived from radar altimetry are usually within ± 3 m of the true elevation, while for surface slopes of $0.7°$ the error increases to around 10 m (Ekholm, Forsberg, and Brozena 1995; Bamber, Ekholm, and Krabill 1998).

Over dry snow packs, the radar signal can penetrate below the surface, giving a potential range error (Martin et al. 1983; Bardel et al. 2002). Attenuation lengths can theoretically be as large as 20 m (Rott and Mätzler 1987), although a figure of a few meters is more common. Techniques to deal with the problem of penetration into the snow pack are based on analysis of the

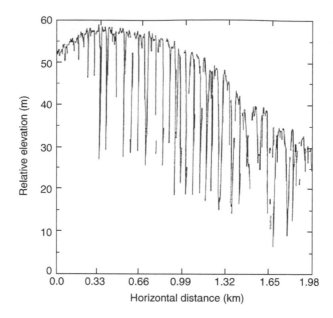

FIGURE 8.7 ATLAS laser altimeter profile of crevasses on Skeidarárjökull, Iceland, 23 September 1991. (Reprinted from Garvin and Williams (1993), with permission of the International Glaciological Society.)

waveform, i.e., of the time-dependence of the returned power (Davis 1993; Ridley and Partington 1988; Yi and Bentley 1994).

Spaceborne radar altimetry samples the topography in a characteriztic "lattice" pattern dictated by the satellite's orbital parameters. Interpolation of the data onto a regular grid can be performed using a number of techniques, such as kriging (Herzfeld, Lingle, and Lee 1993).

Laser profiling is a comparatively new technique for glaciology (Favey et al. 1999). It is still primarily an airborne technique, and hence mainly suited to the study of smaller glaciers. Investigations include those by Echelmeyer et al. (1996), Sapiano, Harrison, and Echelmeyer (1998), Adalgeirsdóttir, Echelmeyer, and Harrison (1998), Thomas et al. (1995), and Kennett and Eiken (1997), with typical height accuracies of 10 to 30 cm and horizontal sample spacings of the order of 1 m. In modern systems, the absolute accuracy of the topographic data is ensured by the use of differential GPS. This performance is good enough to reveal crevasses (Figure 8.7) and even surface meltwater channels (Figure 2.13). Airborne laser profiling of the Greenland ice sheet (Krabill et al., 2000) has shown the same thinning of the marginal regions that was reported by Khvorostovsky, Bobylev, and Johannessen (2003) from analysis of radar altimeter data, although the laser profiling studies did not observe the corresponding thickening of the interior.

Although it was stated earlier that laser profiling is primarily an airborne technique, laser profilers have been placed in space. The first such instrument was the Russian system Balkan, although this provided a height resolution of only 3 m. More recently, the Geoscience Laser Altimetry System (GLAS)* was launched in January 2003 aboard the ICESat satellite. This instrument provides a single nadir-viewing transect with an along-track spacing of about 170 m and a height accuracy of about 10 cm.

Conceptually similar to both radar altimetry and laser profiling, *radio echo-sounding* is also a technique that relies on the timing of short pulses of electromagnetic radiation to deduce the range of a scattering surface from the instrument. Although the primary aim of radio echo-sounding is to measure ice thickness, as discussed in Section 8.4, it yields the range to the surface as a by-product. Because of its wide beam angle it is not a suitable technique for spaceborne use.

The final technique for measuring the surface topography of ice masses is *SAR interferometry*, or InSAR. The essential principles of this technique were described in Section 2.9.4, where it was mentioned that height accuracies of typically a few meters can be achieved. The horizontal resolution is similar to that of the SAR images from which it is generated, and hence usually of the order of 10 m. For a glacier or ice-sheet surface, the two SAR images must be acquired within a few days of each other in order to ensure adequate coherence. The first major opportunity to test the method was provided by the ERS-1/-2 tandem mission in 1995 to 1996, when the satellites ERS-1 and ERS-2, carrying identical SARs, were placed in near identical orbits with a 1-day interval between them. However, ERS satellites placed in 3-day repeat orbits have also proved useful for generating SAR interferograms, as have the longer repeat-period Radarsat and Envisat satellites. Examples of topographic determinations using InSAR are presented by Joughin et al. (1996) for Greenland, and by (Unwin and Wingham 1997) for Austfonna on Svalbard, and Color Figure 8.1 (see color insert following page 108) illustrates a DEM constructed from InSAR.

InSAR is technically difficult to implement, and cannot be described as a wholly predictable technique. Even with a short time interval between the image acquisitions, phase coherence may be lost as a result of a melt event or the deposition of new snow. And although space agencies take great pains to monitor and predict the orbits of satellites, these are subject to perturbations from gravity and the solar wind and the baseline between two SAR observations may differ from the expected value. In practice, the baseline must lie between about 10 m and 1000 m for good results.

8.4 ICE THICKNESS AND BEDROCK TOPOGRAPHY

The thickness of terrestrial ice masses is easier to determine than that of sea ice and freshwater ice. This is because glacier ice is remarkably transparent

---

*http://virl.gsfc.nasa.gov/glas/; http://icesat.gsfc.nasa.gov/intro.html.

to electromagnetic radiation in the MHz to GHz region, which is capable of penetrating kilometers of ice (Bogorodsky, Bentley, and Gudmandsen 1985; Robin, Evans, and Bailey 1969). This provides the rationale for *radio echo-sounding* and *ground-penetrating radar* or *impulse radar*, described in Section 2.8. The basis of the measurement is simply to measure the propagation time for a short radio-frequency pulse to travel to the bedrock and back. For poly-crystalline ice, the propagation speed is around 169 m/μs. However, the speed is significantly greater in the lower-density ice near the surface and this requires corrections to both the apparent thickness and to the position of the point on the bedrock from which scattering is assumed to have occurred (Rees and Donovan 1992). Typical systems for radio echo-sounding are described by Matsuoka et al. (2002d); Christensen et al. (2000). An example of radio echo-sounding data is provided by Bamber and Dowdeswell (1990) for the ice cap Kvitøyjøkulen on Kvitøya, Svalbard (Figure 8.8). An example of surface impulse radar data is given by Jacobel and Bindschadler (1993) (Figure 8.9).

The absorption coefficient* of ice is controlled by its electrical conductivity, and hence primarily by the temperature, though also by the presence of impurities. In addition to providing a measurement of the ice thickness to bedrock, radio echo-sounding also responds to internal structures within the ice (Eisen et al. 2002; Fujita et al. 2002; Paren and Robin 1975), since the radiation can be scattered by ice lenses and ice layers, ash, rock, and dust, chemical precipitates, brine, or temperature stratification. This internal struc-ture can be used to estimate age as a function of depth, and hence to determine the age of buried features such as crevasses (Smith, Lord, and Bentley 2002). Radio echo-sounding has also revealed the presence of subglacial lakes in Antarctica (Drewry 1981). Radio echo-sounding, ground-penetrating radar, and impulse radar are all airborne or surface techniques.

Bedrock topography can be determined directly from knowledge of the surface topography and the ice thickness. It can also be inferred indirectly from the surface topography alone (Budd 1970; Whillans and Johnsen 1983; Fastook, Brecher, and Hughes 1995; Scambos and Haran 2002). This approach is based on a model of ice flow. Budd's model assumes low surface slope, and deformation of ice at the bedrock interface. From the flow model is derived a transfer function that relates the amplitude of surface undulations to that of bedrock undulations. This is maximum (about 0.5) for wavelengths of around 3.3 times the ice thickness, falling to half the maximum value at about 1.5 and 13 times the ice thickness.

8.5 SURFACE TEMPERATURE AND SURFACE MELTING

The surface temperature of a glacier can be determined from thermal infra-red imaging, using, for example, the thermal infrared bands of AVHRR or

---

*The absorption coefficient is the reciprocal of the absorption length, defined in Section 4.2.4.

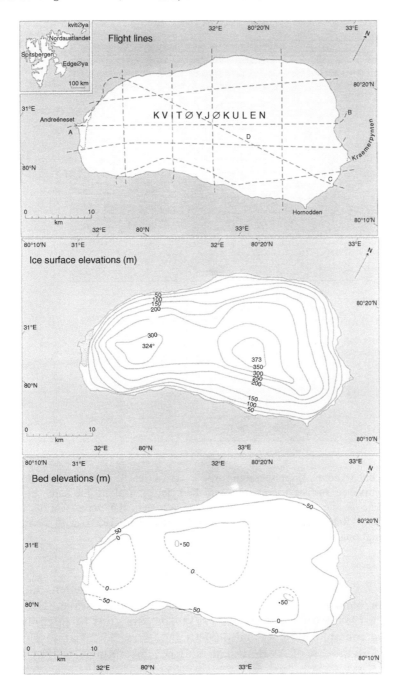

FIGURE 8.8 Typical radio echo-sounding data. (Top) flight lines; (middle) ice surface elevation; (bottom) bed elevation. (Reprinted from Bamber and Dowdeswell (1990) with permission of the International Glaciological Society.)

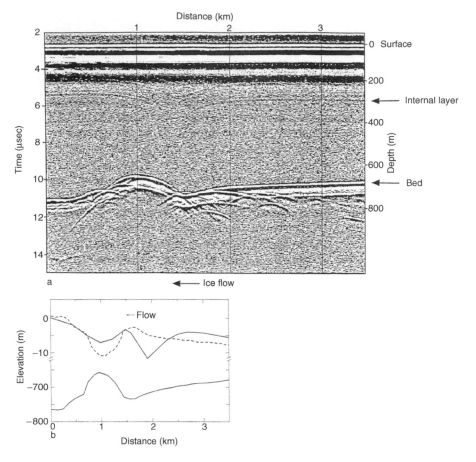

FIGURE 8.9 (Above) Impulse radar profile across Ice Stream D, West Antarctica; (below) surveyed surface topography and bed elevation along the same profile. The dashed line is the hydrostatic surface, showing that the wave feature is not in equilibrium. (Reprinted from Jacobel and Bindschadler (1993), with permission of the International Glaciological Society.)

MODIS for coarse scales (Steffen et al. 1993; Haefliger, Steffen, and Fowler 1993), or band 6 of Landsat TM or ETM+ for finer scales (Orheim and Lucchitta 1988; Pattyn and Decleir 1993). The data must be carefully filtered for cloud contamination* (Comiso 2000) and corrected for atmospheric propagation. In the case of AVHRR data, surface temperatures are often retrieved

---

*One way of discriminating cloud-contaminated pixels in AVHRR imagery is to use band 3 (3.55 to 3.93 μm), since the albedos are dramatically different for snow and cloud in this region (Kidder and Wu 1984). A similar technique can be used with the MODIS channels between 3.5 and 4.0 μm (Ackerman et al. 1998; Riggs, Hall, and Ackerman 1999).

using the split-window technique originally developed for sea surface temperature (Barton et al. 1989), in which the surface temperature T is given by

$$T = a + bT_{11} + cT_{12} \tag{8.4}$$

where T_{11} and T_{12} are the brightness temperatures measured in AVHRR channels 4 (10.5 to 11.5 µm) and 5 (11.5 to 12.5 µm), respectively, and a, b, and c are empirical coefficients determined by regression of surface measurements against the AVHRR data. Radiative transfer modeling using, for example, the LOWTRAN code (Section 2.4.4) can also be used. A modified version of the split-window technique that takes into account the longer atmospheric path for oblique viewing geometries is given by

$$T = a + bT_{11} + cT_{12} + d(T_{11} - T_{12})\sec\theta \tag{8.5}$$

where θ is the scan angle off nadir (Key and Haefliger 1992). For NOAA-11 AVHRR data over the Greenland ice sheet, Haefliger, Steffen, and Fowler (1993) found appropriate coefficients to be $a = -4.26\,K$, $b = 3.47$, $c = -2.47$, $d = -0.14$, giving an RMS error of 0.3 K.

For the Antarctic and Greenland ice sheets, passive microwave data are also useful (being independent of cloud cover) (Das et al. 2002; Shuman and Comiso 2002), although the poor spatial resolution precludes their use for small glaciers. A compensating advantage of the low spatial resolution is the very high temporal resolution available from passive microwave data — typically a couple of days. Passive microwave data are especially useful for indicating the onset of melt, since the emissivity increases dramatically (Zwally and Gloersen 1977) as a result of the increase in effective grain size and the absorption coefficient. The most useful frequency for studying temperature variability is around 37 GHz, since this shows the strongest statistical correlation with air temperature (van der Veen and Jezek 1993; Shuman et al. 1995a). The electromagnetic attenuation length of the order of a meter is similar to the penetration depth of the diurnal thermal variation (Das et al. 2002; Shuman et al. 1995a). The vertical polarization channel is preferred as the emissivity is less strongly dependent on snow parameters in this polarization (Shuman, Alley, and Anandakrishnan 1993). The method can be calibrated against measured or modeled surface temperatures to derive the emissivity; for dry firn, Das et al. (2002) found values of 0.82 ± 0.02 and 0.80 ± 0.02 for the 37 GHz V-polarized channels of the SMMR and SMM/I instruments, respectively. Spatial variations in emissivity as a result of variations in snow/firn conditions are significant (Shuman and Comiso 2002).

Detection of *surface melting* can be carried out using the sharp increase in microwave emissivity referred to above (Abdalati and Steffen 1995; Mote and Anderson 1995; Ridley 1993a, b; Zwally and Fiegles 1994; Ramage and Isacks 2002; Mote et al. 1993). Fahnestock, Abdalati, and Shuman (2002) proposed

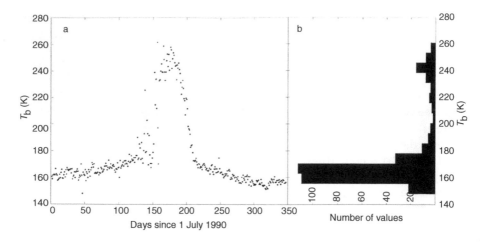

FIGURE 8.10 (a) Temporal variation of 19 GHz H-polarized brightness temperature from a point on the Larsen B ice shelf; (b) histogram of the data shown in (a). The histogram is clearly bimodal with the higher brightness temperatures corresponding to surface melting. (Reprinted from Fahnestock, Abdalati, and Shuman (2002), with permission of the International Glaciological Society.)

an automated technique for deriving a suitable threshold from the annual variation of brightness temperature, based on separating the bimodal histogram of brightness temperature (Figure 8.10). Surface melting can also be identified using the "cross-polarized gradient ratio (XPGR)," defined by Abdalati and Steffen (1995) as

$$XPGR = \frac{T_{19H} - T_{37V}}{T_{19H} + T_{37V}} \tag{8.6}$$

where the quantities on the right-hand side are the horizontally polarized brightness temperature at 19 GHz and the vertically polarized brightness temperature at 37 GHz, respectively. If this ratio exceeds some threshold, it is assumed that liquid water is present near the surface (Fahnestock, Abdalati, and Shuman 2002). The use of diurnal variations in the brightness temperature to identify melting has also been investigated (Ramage and Isacks 2002): the idea is that the night-time emissivity, when the meltwater refreezes, is considerably lower than the daytime emissivity (Mätzler 1987). For 37V data the critical threshold appears to be about 10 K (Ramage and Isacks 2002).

Surface melting can also be detected using scatterometer measurements (Wismann 2000) (Figure 8.11), since the backscattering coefficient decreases sharply with increasing water content. In contrast to the use of passive microwave radiometry, the technique offers higher spatial resolution and, by virtue of its longer wavelength, greater penetration into the surface layer.

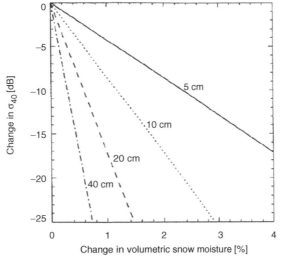

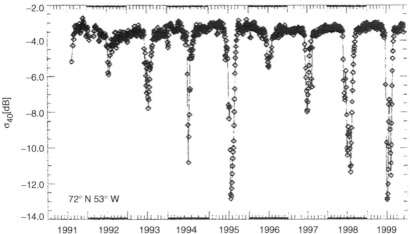

FIGURE 8.11 (Above) Modeled change in CVV backscattering coefficient at an incidence angle of 40° as a function of snow moisture and thickness of the wet-snow layer; (below) time-series of scatterometer measurements from a location at the boundary between the wet-snow and percolation zones on the Greenland ice sheet. The spikes correspond to melting events. (Reproduced from Wismann (2000), figures 2 and 3. Copyright 2000 IEEE.)

8.6 ACCUMULATION RATE

Passive microwave radiometer data can also be used to estimate accumulation rates in the dry-snow zone, either using empirical relationships between emissivity, grain size, and accumulation rate (Zwally 1977; Zwally and Giovinetto 1995; Abdalati and Steffen 1998), or by observing correlations between the time-variation of microwave brightness temperature and the layering structure observed within snow pits (Shuman et al. 1995b; Alley et al. 1997). The former technique can be enhanced by combining data from radar altimetry (Davis 1995). Areas of unusually large grain size in central Antarctica (arising essentially from very low accumulation rates) have been detected in a similar manner through their unusually low brightness temperatures (Surdyk 2002).

8.7   SURFACE FACIES

The division of the surface of a glacier into different zones or facies was discussed in Section 4.5.1. The ability of a remote sensing method to distinguish these facies depends on its ability to recognize the different surface types (dry snow, wet snow, bare ice) at different times of year, and also (in the case of radar methods) to identify the presence of subsurface features.

VIR imagery has been applied to the detection of surface facies with some success. Imagery acquired at the end of the ablation season allows some discrimination of surface facies (Hall et al. 1987; Williams, Hall, and Benson 1991). This is rather a narrow time window, which can make the acquisition of suitable imagery difficult. Three zones can usually be distinguished, identified as snow, slush, and ice zones by Williams, Hall, and Benson (1991) and Hall et al. (1987). The snow zone encompasses both the wet-snow and percolation zones (Williams, Hall, and Benson 1991), while the slush zone is probably a transitional region that consists at least partially of wet snow. Difficulties in delineating these zones are caused by fresh snowfall, which can obscure the glacier surface (Hall et al. 1987; Williams, Hall, and Benson 1991), and by the presence of debris on the surface, which can reduce the contrast between bare ice and surrounding moraine. On the other hand, the presence of debris can enhance the contrast between the ablation and accumulation areas (König, Winther, and Isaksson 2001). Most approaches to classification of VIR imagery of glacier surfaces are based on tonal differences, although the use of texture measures has also been investigated and found to be effective (Bishop et al. 1998).

The snow line can be discriminated as the lower boundary of the slush zone, more precisely if the imagery is corrected for topographic variations using a DEM (Parrot et al. 1993). Early work on snowline detection was reported by Krimmel and Meier (1975). More recent work (Seidel et al. 1997) has emphasized the idea of the snow line as a statistical concept. The equilibrium line cannot be detected directly, but can be inferred from the position of the snow line at the end of the ablation season provided that there is no significant quantity of superimposed ice (Winther 1993).

SAR imagery offers substantially greater scope than VIR imagery for the discrimination of glacier facies. It offers the usual advantages of being independent of daylight and cloud-free conditions, but the interactions between the microwave radiation and the surface and subsurface of the glacier are also more diverse than the corresponding interactions for VIR imagery. Most SAR imagery from glaciers that has been analyzed to date has been "single-parameter" data, i.e., a single frequency and polarization state, which increases the possibility of ambiguity in identifying the cause of a particular backscattering phenomenon. Nevertheless, early analysis of SAR imagery of ice sheets (Rott and Mätzler 1987; Bindschadler, Jezek, and Crawford 1987; Rees 1988) demonstrated its ability to identify many features on and adjacent to the ice sheet. Fahnestock and Bindschadler (1993) drew attention to SAR's sensitivity to surface conditions and potential ability to map surface facies, and detailed investigation of this ability continued through the 1990s, stimulated especially by the launch of ERS-1 in 1991.

If a SAR image is available only for a single date, the most useful time of year is winter when the snow cover is most likely to be dry and hence transparent to microwave radiation (Ahlnäs et al. 1992). This is opposite to the situation for visible imagery, where summer imagery is preferred, and gives a much longer time window for data acquisition (König, Winther, and Isaksson 2001). Based on the discussion in Section 4.5.4, one would usually expect to see low backscatter from the bare ice zone (smooth surface and low volume scatter), low but patchy backscatter from the wet-snow zone (absorption), high backscatter in the percolation zone (scattering from ice lenses and layers), and low backscatter from the dry-snow zone (Bindschadler, Jezek, and Crawford 1987; Rott and Mätzler 1987; Partington 1998; Rott, Sturm, and Miller 1992; Jezek 1992). Bindschadler, Fahnestock, and Kwok (1992) constructed a mosaic of 18 winter ERS-1 SAR images of west Greenland, terrain-corrected using a DEM. They presented only a qualitative analysis of the backscattering coefficient data, but showed that all four zones could be discriminated and showed a strong correlation with elevation. Features on scales less than 1 km are usually topographically controlled variations in melt features (Vornberger and Bindschadler 1992). Bindschadler, Fahnestock, and Kwok (1992) noted that the Jakobshavn Isbrae glacier exhibited high backscatter, attributed to its heavy crevassing. Dowdeswell, Rees, and Diament (1993) and Bindschadler, Fahnestock, and Kwok (1992) also noted the ability of winter imagery (and summer imagery to a lesser extent) to reveal ice divides.

A more extensive winter mosaic of Greenland was constructed by Fahnestock et al. (1993). Minor variations of backscatter coefficient within the dry-snow zone may be associated with variations in accumulation rate (Jezek 1992): because snow grains grow with age, areas with lower accumulation rates will tend to have larger grains near the surface and hence produce higher volume scattering. Using SAR data from Antarctica, Bardel et al. (2002) confirmed the general pattern that the ablation area usually exhibits low backscatter in winter imagery and the accumulation area high backscatter.

Summer SAR imagery is more difficult to interpret than winter imagery since, as discussed in Section 4.5.4, surface topography, surface roughness, and water content all play a role in determining the backscatter.* The ambiguities can be reduced by applying an altitude mask to the data (Rau and Braun 2002) or by using image texture in addition to backscatter coefficient as a discriminant (Rees and Lin 1993). However, the most useful approach with single-parameter SAR[†] is undoubtedly through the use of multitemporal data. Early work was described by Dowdeswell, Rees, and Diament (1993) and Rees, Dowdeswell, and Diament (1995) using data from Austfonna, Svalbard. The general pattern, confirmed by Braun et al. (2000) in observations of King George Island, Antarctica, and by Ramage, Isacks, and Miller (2000) for the Juneau Icefield, Alaska (Figure 8.12), is for high backscatter everywhere in winter unless a dry-snow zone is present, decreasing in the wet-snow zone by around 10 dB at the time of the spring melt, followed by a spatially and temporally variable period during the summer. The multitemporal approach has been explored in detail by Partington (1998), who used multitemporal false-color composites (red = early summer, green = late summer, blue = winter) to identify zones. The dry-snow zone appears as dark gray (low backscatter throughout the year), the percolation and wet-snow zones are white or light gray in the upper region, purple in the middle, and blue at the bottom (this depends on which areas are melting and which are freezing), while the bare ice zone is greenish. Although it is valuable to have SAR images throughout the year, most information is obtained from two measurements: the winter backscatter, and the difference between the end-of-ablation and winter backscatters.

Because the zonation of SAR images does not necessarily match Benson's glaciological facies, the term "radar glacier zones" has been proposed (Forster, Isacks, and Das 1996; Smith et al. 1997). The proposed zones are listed here (Braun et al. 2000), although not all of these have been observed by all authors, and their radar characteristics are in some cases still uncertain:

1. The dry-snow radar zone, in which melting never occurs and backscatter values, dominated by volume scattering, are low (typically −10 to −20 dB) (Fahnestock et al. 1993; Partington 1998) throughout the year. This zone occurs only in Antarctica and Greenland, and on some Svalbard and Alaskan glaciers at high altitude.
2. The frozen-percolation radar zone, with frequent ice layers, large grain sizes, and high backscatter (typically −3 to −8 dB) during the

---

*Indeed, backscatter modeling has been used by a number of authors to estimate these physical properties from radar data. An example is given by Smith et al. (1997), who found that the geometric optics model provided a good description of wet snow in the melt zone, provided that the snow had not metamorphosed.

[†]Some preliminary work has indicated that multipolarization (Forster and Isacks 1994; Rott 1994; Rott and Davies 1993; Shi and Dozier 1993; Rott, Floricioiu, and Siegel 1997) and multifrequency (Rott and Davies 1993; Rott, Floricioiu, and Siegel 1997) data are promising for resolving some of the ambiguities in single-parameter radar images of glaciers, but this work has still not been fully developed.

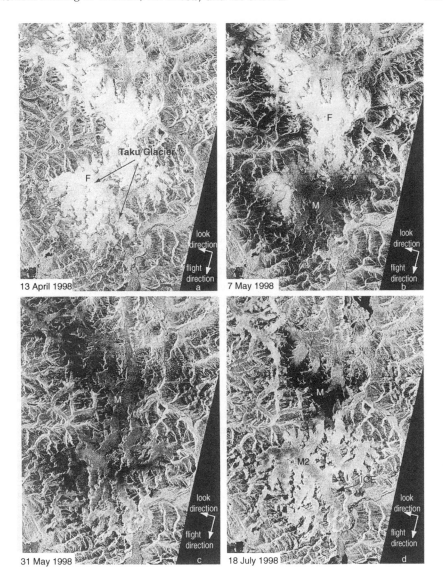

FIGURE 8.12 Time-series of RADARSAT SAR images of the Juneau Icefield, Alaska. F: frozen; M: initial melt; M2: phase 2 melt; ICE: bare ice. (SAR data Copyright CSA 1998. Reprinted from Ramage, Isacks, and Miller (2000), with permission of the International Glaciological Society.)

winter months (Fahnestock et al. 1993; Bindschadler and Vornberger 1992). During the summer the surface is wet and the backscatter is substantially reduced.

3. The wet-snow radar zone, with winter backscatter intermediate (typically −14 to −22 dB) between the dry-snow and percolation

zones (Fahnestock et al. 1993; Bindschadler and Vornberger 1992). Discrimination between zones 2 and 3 may be ambiguous, and it may be necessary to regard them as a single continuous transition (Partington 1998; Dowdeswell, Rees, and Diament 1993).

4. The phase 2 (P2) melt radar zone, a zone of enhanced backscatter that can appear briefly in winter and has been attributed (Smith et al. 1997) to metamorphosed and roughened melting snow (typically -4 to -8 dB).

5. The superimposed ice radar zone (SIZ). This does not generally occur on temperate glaciers in late summer, since the snow line is nearly coincident with the equilibrium line (König, Winther, and Isaksson 2001). Where it does occur, it forms by a different mechanism from glacier ice, and has different physical properties (Koerner 1970). It has been suggested (Bindschadler and Vornberger 1992) that it is virtually impossible to distinguish the SIZ from glacier ice in radar imagery. However, Marshall, Rees, and Dowdeswell (1995) reported successful detection on Ayerbreen, Svalbard, which they attributed to substantial differences in surface roughness in late-summer images. This was also investigated by König et al. (2002) for two glaciers in Svalbard. These authors found that winter and spring images of Kongsvegen were clearly differentiated into three zones of increasing backscatter, and on the basis of fieldwork they established that these represented glacier ice (-12 ± 1 dB), summer-formed superimposed ice (-6 ± 2 dB), and firn (-1 ± 1 dB) (Figure 8.13). They suggested that the difference is caused by differences in the size distribution of included air bubbles. On the smaller glacier Midre Lovénbreen they observed a bare ice facies (-11 ± 1 dB) and a zone of intermediate backscatter (-5 ± 1 dB) corresponding to summer-formed super-imposed ice (Figure 8.14) adjacent to the bare ice zone, and distinguished from it by a lower backscatter coefficient as a result of its smoother surface.

6. The *bare glacier ice zone*, with a comparatively strong backscatter from the ice surface (typically -10 to -13 dB).

Can the late-summer snow line be detected in radar imagery, and is it coincident with the equilibrium line? Positive evidence has been provided by Braun et al. (2000), Rees, Dowdeswell, and Diament (1995), Engeset and Weydahl (1998), Hall et al. (2000), König, Winther, and Isaksson (2001), and Partington (1998), although some ambiguities remain (Strozzi, Wegmuller, and Mätzler 1999) and the presence of a superimposed ice zone and the exposure of last year's firn introduce uncertainties. Demuth and Pietroniro (1999) analyzed late-summer Radarsat imagery from Peyto glacier, Canada, and found that the boundary between bare ice and firn could be easily identified and traced manually or using a simple minimum-distance-to-means decision rule based on training data. (The data were corrected for incidence angle effects.) Bare ice was

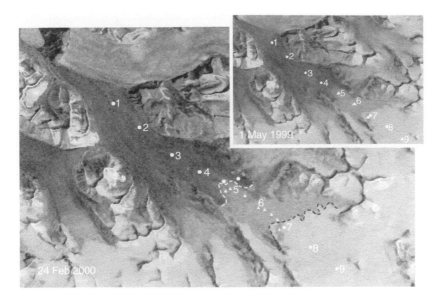

FIGURE 8.13 ERS-2 SAR images of Kongsvegen, Svalbard. Circles (numbered 1 to 4) are areas of glacier ice; triangles (5 to 7) are areas of superimposed ice; squares (8 and 9) are areas of firn. (SAR images Copyright ESA 2000. Reprinted from König et al. (2002), with permission of the International Glaciological Society.)

brighter than firn by about 7 dB. However, there remains some doubt on this point. Analysis of 8 years of radar data from the Kongsvegen glacier by Engeset (2000) and Engeset et al. (2002) suggests that the correspondence of a boundary in the radar image to the position of the equilibrium line is coincidental, and that the boundary actually corresponds to the firn line (König, Winther, and Isaksson 2001). A complementary technique has been investigated by Kelly (2002) based on repeat-pass interferometric SAR. The idea here is not to generate interferograms but merely to use the coherence between the images as an indicator of changing surface conditions: the transient snow line is characterized by a low coherence over a period of 24 hours bounded by more coherent regions up-glacier and down-glacier. The ERS tandem mission (with the ERS-2 satellite following 1 day behind ERS-1) proved suitable for this investigation. The results were promising, and suggested that the transient snow line can be detected by examining the difference between a winter and a summer coherence image. Further investigation of this possibility would require another mission similar to the ERS tandem phase to take place.

8.8 ICE MOTION

There are a number of approaches to the determination of ice motion from remotely sensed data, both direct and indirect. Indirectly, motion can be

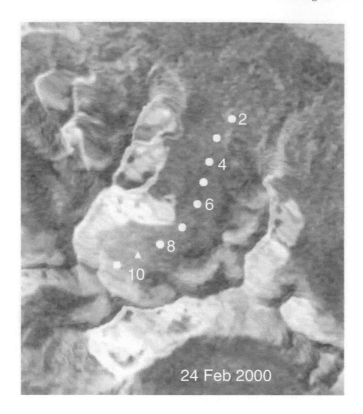

FIGURE 8.14 Winter ERS-2 SAR image of Midre Lovénbreen, Svalbard, showing low backscatter from glacier ice areas (circles; numbers 2 to 8) and intermediate backscatter from superimposed ice (triangle; unnumbered). (Image data Copyright ESA 2000. Reprinted from König et al. (2002), with permission of the International Glaciological Society.)

estimated from a detailed surface topography and assumptions about the ice flow. However, the most obvious direct approach is through the identification of surface features such as foliation, crevasses, and moraines, and tracking their motion in a time-series of images (Hambrey and Dowdeswell 1994). The features themselves can be identified manually (Lucchitta et al. 1993; Lefauconnier, Hagen, and Rudant 1994; Ferrigno et al. 1993) or by using a cross-correlation technique similar to that described for sea ice dynamics (see Section 6.6) (e.g., Scambos and Bindschadler 1993; Scambos et al. 1992). Clearly, good image-to-image registration is necessary for this technique. If enough stable ice-free features (e.g., nunataks (Figure 1.6) or exposed rock surrounding a glacier) are visible in the images, these can be used as control points. This approach is usually satisfactory for glaciers and the peripheries of ice sheets but not for the interiors of ice sheets, where such features are not usually present. An approach that has been adopted in such cases is to coregister large-scale features (> 1 km) that are controlled by the bedrock topography (Scambos et al. 1992; Whillans and Tseng 1995). Accuracies as good as 6 m have been reported for Landsat TM imagery. These techniques have provided new insights into the dynamics of ice streams (Bindschadler 1998). Although feature-tracking is primarily applied to VIR imagery, it can also be carried out using SAR imagery (Fahnestock et al. 1993; Lucchitta, Rosanova, and Mullins 1995; Murray et al. 2002). Here the technique is sometimes referred to as intensity tracking. Speckle tracking, in which the features are the speckle pattern of the radar image, is also possible (Michel and Rignot 1999; Gray, Mattar, and Vachon 1998).

Motion of the ice surface can also be determined using SAR interferometry (InSAR). As has been noted in Section 2.9.4, the interferogram generated from a pair of SAR images combines information about both topography and motion, and in this case it is clearly necessary to separate these two effects. If the baseline is very short the effect of topography is minimized, or three images can be used instead of two (Kwok and Fahnestock 1996; Joughin, Kwok, and Fahnestock 1998). Alternatively, a DEM can be used to remove the topographic component from the data. Phase ambiguity in the data means that at least one reference point is needed to constrain the data: either a known fixed point (e.g., exposed rock) or a surveyed point of known velocity. The typical accuracy of displacement measurements is 1 mm at C band (Bindschadler 1998). In fact, InSAR can only measure the component of velocity that is parallel to the radar view angle (see Color Figure 8.1 following page 108). It is usually assumed that the direction of the flow velocity is along the steepest surface slope (Joughin, Kwok, and Fahnestock 1996, 1998) or parallel to valley walls (Fatland and Lingle 1998), but one can also use more than one pair of images with different look directions (ascending and descending passes). Like all InSAR measurements, velocity determination depends on phase coherence between the images. If the time interval is long for repeat-pass interferometry, this can be a problem. The Radarsat Antarctic Mapping Mission AMM-1 (Jezek 1999) had a repeat period of 24 days, which is very long by InSAR standards. Coherence was generally maintained only for areas with low accumulation rates, less than about 15 cm/yr, and in areas not strongly affected by katabatic winds or dynamic atmospheric conditions in general (Joughin 2002; Wunderle and Schmidt 1997). For the Radarsat repeat period and imaging geometry, the accuracy in velocity determination is good for low speeds (less than about 100 m/yr), but higher speeds cause difficulty in unwrapping the phase (Joughin 2002). (Velocities on an ice shelf can reach 1000 m/yr or more (Young and Hyland 2002).) Joughin (2002) has developed a technique that combines both InSAR and speckle tracking to cope with higher flow speeds.

InSAR has also been used to monitor the flexure of ice shelves (of the order of a meter) and demonstrate that it is consistent with physical theory, yielding estimates of the position of the grounding line and the elastic modulus of the ice shelf (Schmeltz, Rignot, and MacAyeal 2002; Gray et al. 2002). Laser ranging will also have the potential to provide such information (Padman et al. 2002).

An interesting example of InSAR motion measurement is described by Gudmundsson et al. (2002) who studied the infilling of the ice depression caused by a subglacial eruption of the Gjálp volcano under Vatnajökull, Iceland in 1996.

8.9 MASS BALANCE

The most obvious way of determining the mass balance of a glacier is to make repeated measurements of its surface topography, using airborne stereophotography (Andreassen, Elvehøy, and Kjøllmoen 2002) or laser profiling (Favey et al. 1999) for small glaciers. The spatial distribution of the mass balance can be mapped by combining time-difference DEMs with the surface velocity field (Hubbard et al. 2000). For the central regions of Antarctica and Greenland radar altimetry or airborne laser profiling can be used, though with difficulty (Wingham et al. 1998; Filin and Csathó 2002). Repeated airborne laser profiling surveys have shown that the Greenland ice sheet is stable in the center, though thinning at up to 1 m/yr toward the coast (Krabill et al., 2000). This technique has also been used in Antarctica, Alaska, and the Swiss Alps (Filin and Csathó 2002).

For glaciers on the peripheries of ice sheets the situation is more difficult (Rignot 2002). Rignot (2002) adopted the following approach. The output ice flux is estimated from the product of ice shelf velocity (from ERS InSAR) and the cross-sectional area, the thickness being derived from a suitable DEM (Bamber and Bindschadler 1997) and the assumption that the shelf is in hydrostatic equilibrium. The input ice flux is calculated by integrating the accumulation rate (Giovinetto and Zwally 2000; Vaughan et al. 1999) over the delineated drainage basin. A correction is applied for basal melting. This work showed that the position of the grounding line of several major glaciers had previously been substantially misestimated. Most of the glaciers are in fact more or less in balance. A similar approach was successfully adopted by Rignot et al. (2000) for Nioghalvfjerdsbrae on Greenland.

Mass balance can also be estimated indirectly from changes in the distribution of the surface zones, e.g., expansion of the ablation area is indicative of negative mass balance. This can be done using VIR imagery (Krimmel and Meier 1975) or SAR. Mass balance can be inferred indirectly from the equilibrium line elevation (ELA), provided that the relationship between the two variables has been calibrated. As was noted earlier, the equilibrium line often coincides more or less with the late summer snow line, and techniques for identifying this were discussed in Section 8.7.

9 Remote Sensing of Icebergs

9.1 INTRODUCTION

The detection, monitoring, and measuring of icebergs represents one of the more difficult challenges in cryospheric remote sensing. The significance of icebergs is great, as discussed in Chapter 1. In the Antarctic, most of the mass lost from the ice sheet takes place as basal melting and iceberg calving from ice shelves and glacier tongues (Jacobs et al. 1992), and there are estimated to be over 200,000 icebergs south of the Antarctic Convergence (Williams, Rees, and Young 1999). In the northern hemisphere, most icebergs calve from the Greenland ice sheet and the archipelagos of the Barents Sea, and again, an understanding of the quantity of ice discharged is important in assessing the mass balance of the ice sheets and glaciers that produce them. The role of icebergs as a hazard to shipping and offshore facilities was also discussed in Chapter 1. Yet most icebergs are small and mobile, and hence difficult to identify and to track.

9.2 DETECTION AND MONITORING OF ICEBERGS

9.2.1 VISIBLE AND NEAR-INFRARED OBSERVATIONS

The high radiometric contrast between an iceberg and the surrounding ocean facilitates its detection. Conventional aerial photography can be used for detecting icebergs and analyzing their size distribution (Vefsnmo et al. 1989; Løvås, Spring, and Holm 1993), and has been so used for several decades (Rossiter et al., 1995). When operated from conventional aircraft, if offers very high spatial resolutions (typically a few centimeters) but correspondingly limited spatial coverage (a few kilometers). In a typical application, Vefsnmo et al. (1989) used a mapping camera from an altitude of 914 m during the IDAP program to photograph 135 icebergs in the Barents Sea, south east and south of Spitsbergen. Løvås, Spring, and Holm (1993) describe a more extensive study of iceberg distribution in the same region, covering the period from 1988 to 1992. The typical sizes of icebergs reported in this study were 15 m freeboard, 85 m length, and 150,000 tonnes mass.

The principal limitation of aerial photography is its spatial coverage. Satellite imagery provides coverage of much larger areas, although at the expense of reduced spatial resolution, and has been used to identify and track the motion of icebergs since the late 1960s. For example, Swithinbank, McClain, and Little (1977) described the tracking of 14 Antarctic icebergs over a period of up to 9 years. The use of wide-swath visible imagery, such as NOAA AVHRR, MODIS, or DMSP OLS, is favorable because of its high temporal resolution (Ferrigno and Gould 1987; Keys, Jacobs, and Barnett 1990). If the iceberg is large enough, such imagery can provide sufficient spatial resolution to monitor changes in its morphology as well as tracking its trajectory. An example of this is given by Argentina (1992) who used AVHRR imagery to track the 13,000 km² iceberg that calved from the Filchner ice shelf in Antarctica in 1986. This iceberg split into three large grounded ice islands almost immediately, and one of these became dislodged in early 1990 and began to drift northward, covering about 2000 km in 19 months. Similar mean speeds of a few centimeters per second have been reported by other satellite tracking of large icebergs, for example Komyshenets and Leont'yev (1989), who tracked the large (5900 km²) berg that calved from the Larsen ice shelf in 1986.

Narrower-swath, higher-resolution imagery, such as Landsat or SPOT, is also valuable for characterizing iceberg distributions (Vinje 1989). Kloster and Spring (1993) used Landsat Thematic Mapper (TM) and Multispectral Scanner (MSS) imagery, and SPOT HRV imagery, to investigate the distribution of smaller icebergs that had calved from glaciers of the Franz Josef Land archipelago in the Barents Sea. Of the three, they found the TM imagery to be best suited to identification of icebergs, principally as a result of its greater areal coverage than SPOT HRV and higher spatial resolution than MSS. The single most useful spectral band of the TM data was band 4 (0.76 to 0.90 μm). Icebergs down to 2–3 pixels in size (1800 to 2700 m²) could be detected, based on one or more of these "signatures:"

1. Detection of a shadow
2. Strong reflection from the sun-facing part of the iceberg
3. Texture or albedo characteriztics (in the case of large icebergs)
4. Detection of a lead or wake

The first two of these imply that the optimum time for detecting icebergs using VIR data is March to May in the northern hemisphere (Sandven, Kloster, and Johannessen 1991). The usefulness of each of these signatures depends on a number of environmental parameters, especially the solar altitude (which controls the length of the shadow and the contrast against the background) and the nature of the surrounding material. Grounded icebergs are often easy to detect if they are surrounded by pack ice in an area of strong tidal currents (Johannessen, Sandven, and Kloster 1991; Vefsnmo et al. 1989; Sandven, Kloster, and Johannessen 1991), since the relative motion of the pack ice to the iceberg leaves a wake of open water or crushed ice (Figure 9.1).

In general, icebergs are most easily detected in early spring (March to May in the Arctic), when they are surrounded by smooth sea ice and cast long shadows as a result of the low solar elevation (Figure 9.2) (Sandven, Kloster,

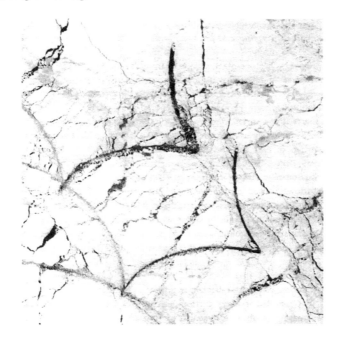

FIGURE 9.1 SPOT image (10 km × 10 km) of Spitsbergenbanket, 11 April 1988, showing wakes in pack ice left by grounded icebergs. (Reproduced from Johannessen, Sandven, and Kloster 1991.)

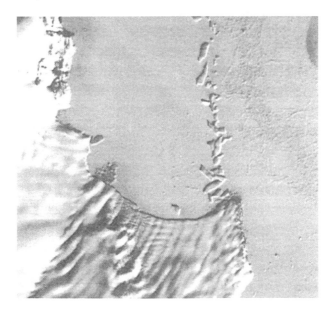

FIGURE 9.2 Renown glacier front, Franz Josef Land, in a TM image from 9 April 1988, showing the glacier front and icebergs. The appearance of the icebergs is enhanced by the low sun angle. (Reproduced from Kloster and Spring 1993.)

and Johannessen 1991). The length of the shadow can be used to estimate the iceberg's freeboard. By contrast, mapping of glacier fronts is most usefully carried out using late summer imagery, in which open water in front of the ice provides a strong contrast.

Airborne VIR scanners are also available and applicable to the monitoring of icebergs, although this possibility does not appear to have been extensively explored.

9.2.2 Passive Microwave Radiometry

The major disadvantage of VIR imagery for tracking icebergs is obscuration by cloud cover. For sufficiently large icebergs, spaceborne passive microwave radiometry offers a solution to this problem despite its low spatial resolution. For example, Hawkins et al. (1993) used a combination of AVHRR, OLS, and SSM/I imagery to track the Antarctic tabular iceberg A24 (70×80 km; part of the 1986 calving from the Filchner ice shelf) and to observe its eventual decay in April 1992. As with the AVHRR imagery, the wide swath of the SSM/I imagery provides a high temporal resolution. On the other hand, the poor spatial resolution achievable by spaceborne passive microwave systems limits its usefulness to very large icebergs. Airborne systems can provide much higher spatial resolutions. For example, the Airborne Imaging Microwave Radiometer (AIMR), developed in 1989, operates at 37 and 90 GHz and provides a resolution of 84 and 35 m, respectively, over a swath of 7 km, when flying at a height of 2000 m (Rossiter et al., 1995).

9.2.3 Synthetic Aperture Radar

Higher spatial resolution, together with the ability to generate images through cloud cover and at night, is provided by SAR and other imaging radar systems. Airborne real aperture and synthetic aperture radars (SLARs and SARs) have been used for iceberg detection since the early 1970s and early 1980s, respectively (Rossiter et al., 1995; Kirby 1982), and spaceborne SAR systems have been continuously available since the launch of ERS-1 in 1991. The various interaction mechanisms between microwave radiation and icebergs have been discussed by Willis et al. (1996) and Johannessen, Sandven, and Kloster (1991). In addition to surface scattering and volume scattering from the iceberg, these include the "double-bounce" (dihedral scattering) mechanism, in which radiation is scattered first from the vertical wall of an iceberg and then from the horizontal sea surface, or vice versa. This mechanism accounts for many of the strong localized responses from the edges of ice islands (Jeffries and Sackinger 1990) and the "point-target" response of very small icebergs. Icebergs can also cast radar shadows (Figure 9.3), and unlike optical shadows, no radiation whatsoever is received from these regions (i.e.,

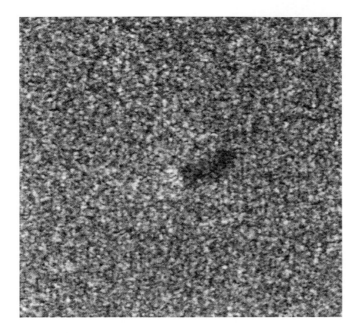

FIGURE 9.3 Enlargement of an ERS-1 SAR image showing a small iceberg and its radar shadow. (Reproduced from Willis et al. (1996), with permission of Taylor & Francis plc.)

they are completely dark). If the spatial resolution is high enough, and the incidence angle* of the radar is large enough, this phenomenon can be used to estimate the three-dimensional geometry of the iceberg (Larson et al. 1978). The wakes left by icebergs, either in surrounding water or in pack ice that is motion relative to the iceberg, can also provide a characteristic signature in a manner similar to that discussed for visible-wavelength imagery (Larson et al. 1978; Johannessen, Sandven, and Kloster 1991; Sandven, Kloster, and Johannessen 1991) (Figure 9.4). Other possible effects on the surrounding sea surface include lee shadowing, giving a calmer sea surface, and damping of waves by meltwater runoff. Table 9.1, adapted from Willis et al. (1996), summarizes the potential of these various imaging mechanisms for the detection of icebergs.

Early demonstrations of the potential of SAR to study icebergs were given by Larson et al. (1978) and Kirby and Lowry (1979), as a result of which it was apparent that higher-frequency systems (e.g., X band) would be preferable to

---

*For this and other reasons, imaging radars should have large incidence angles to be optimal for iceberg detection. Early spaceborne SAR systems had rather small incidence angles (typically around 20 to 30°). However, Radarsat and later spaceborne SAR systems have provided larger incidence angles.

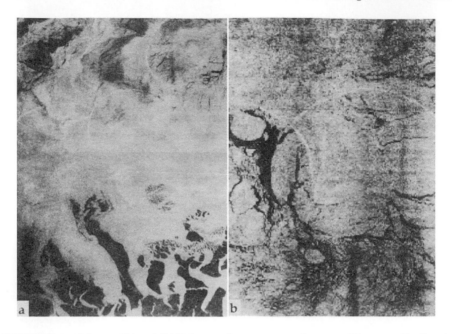

FIGURE 9.4 Airborne C-band SAR images from near northeastern Bjørnøya, Svalbard, 23 February 1989. (b) is at a larger scale than (a). The images show iceberg trajectories as roughly cycloidal loops. (Adapted from Sandven, Kloster, and Johannessen (1991) with permission of the International Society for Offshore and Polar Engineering.)

TABLE 9.1
Summary of Possible Detection Techniques for Icebergs in SAR Imagery

Imaging Mechanism	Possible Detection Mechanisms	Applicability
Double-bounce scattering	Bright point-target detection	Generally useful
Shadowing	Dark point-target detection	Generally useful; better at high incidence angle
Effect on surrounding sea surface	Segmentation, Hough transform, Fourier transform	Generally useful
Surface texture	Image segmentation	Large icebergs
Shape	Image segmentation	Large icebergs
Wakes	Segmentation, Hough transform	Unknown
Volume scattering	Mulitfrequency or polarimetric data	Multiparameter SAR
Bottom reflection	Correlation methods	L band only

Adapted from Willis et al. (1996).

the longer-wavelength L-band systems as a result of the greater attenuation length at low frequency.* A 100% detection rate for large icebergs was reported by Lowry and Miller (1983) using an airborne X-band SAR, though they did not report the spatial resolution of their system. They found that icebergs smaller than about 20 m were essentially undetectable. Livingstone et al. (1983), also using an airborne X-band SAR, reported that the smallest detectable iceberg was typically four to six times the spatial resolution, presumably as a result of image speckle. Early attempts at identifying techniques for delineating icebergs in digital SAR imagery, against a background of sea ice, were reported by Kirby (1982) but were not particularly successful, although the author did point out the value of median filtering (Section 3.3.2) as a technique for suppressing image speckle. The first comprehensive study of the relationship between iceberg detectability and size appears to have been carried out by Willis et al. (1996) using ERS-1 SAR imagery. They used a technique based on the identification of adjacent bright and dark returns, corresponding to the iceberg and its wake, and found that essentially all icebergs larger than 120 m, and about half of all icebergs between 15 and 60 m, could be detected against a background of calm sea water. A disadvantage of their technique is its proneness to false alarms, i.e., the detection of nonexistent icebergs, and its inability to provide an estimate of the size of an iceberg unless it is greater than about 220 m.

A different approach to the detection of icebergs was adopted by Williams, Rees, and Young (1999), who adapted an image segmentation technique originally described by Sephton et al. (1994). In this approach, a pixel-bonding process is used to delineate the edges of an iceberg, followed by the application of an edge-guided segmentation process to separate the iceberg from its background. The algorithm was tested using ERS-1 data. In order to reduce the problem of speckle corrupting the edges of the image of an iceberg, the SAR data were first smoothed to a resolution of 100 m. At this resolution, the algorithm was capable of detecting all icebergs of at least 6 pixels ($6 \times 10^4 \, \mathrm{m}^2$) in size, although the larger bergs tended to be oversegmented (Figure 9.5).

A number of authors (Lowry and Miller 1983; Rossiter et al. 1984; Jeffries and Sackinger 1990; Shokr, Ramsay, and Falkingham 1992) have shown that the detectibility of icebergs is decreased by higher sea states, as a result of the increasing backscatter from rougher seas. Indeed, if the sea state is sufficiently high, the radar return from the sea can be greater than that from the iceberg (Gray, Livingstone, and Hawkins 1982).

Jeffries and Sackinger (1990) investigated the potential of SAR imagery for the detection and characterization of ice islands using airborne X-band SAR to image the ice island "Hobson's Choice" in the Canadian Archipelago.

---

*On the other hand, L-band SAR imagery of icebergs can exhibit bright "ghost" images on the far-range side of the iceberg, corresponding to reflection of the penetrating radiation from the ice–water interface (Gray and Arsenault 1991). Under some circumstances, these "ghosts" may enhance the detectability of icebergs.

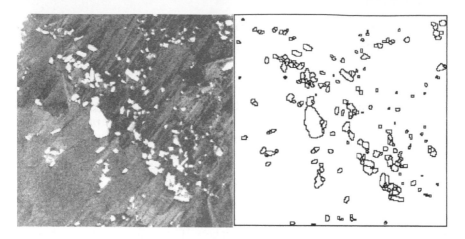

FIGURE 9.5 Iceberg detection by segmentation (Williams, Rees, and Young 1999). (Left) Part of an ERS-1 SAR image (23 × 23 km) off the Antarctic coast, north east of Davis station. The image has been smoothed to a resolution of 100 m to reduce the effects of speckle. (Right) The result of applying a segmentation algorithm to this image to delineate the icebergs in it. (Reproduced with permission of Taylor & Francis plc; SAR image copyright ESA 1993.)

Figure 9.6 illustrates the inhomogeneous nature of this ice island, which consists of shelf ice, sea ice that was attached to the shelf ice at the time of calving and has remained so, and sea ice that has subsequently become attached to the island. In other words, the ice island consists of a mixture of freshwater and saline ice, the latter giving a significantly lower backscatter as a result of the reduced penetration of the radiation into the more electrically conductive material. The image texture of these areas is also quite different, the freshwater shelf ice exhibiting a striated pattern characteristic of the undulations in the Ward Hunt ice shelf from which it calved. Figure 9.6 also illustrates the difficulty of detecting an ice island of this type against a background of pack ice, a difficulty also identified by Vinje (1989), who demonstrated instances in which tabular icebergs as large as 1600 m could not be identified. As Jeffries and Sackinger (1990) acknowledge, detection of ice islands in SAR imagery may benefit considerably from prior knowledge.

The suitability of spaceborne SAR, and indeed any spaceborne system, for monitoring icebergs depends on many factors. These embrace the spatial resolution and the reliability of the detection mechanism, but the temporal resolution and the data turnaround time are also major factors in any operational system. The early spaceborne SAR systems (Seasat, ERS-1) were narrow-swath instruments with correspondingly long revisit times for any specific location. More recently, multimode spaceborne SARs (Radarsat, Envisat) have been developed and deployed. These can be operated in wide-swath modes with correspondingly improved temporal resolution, although the

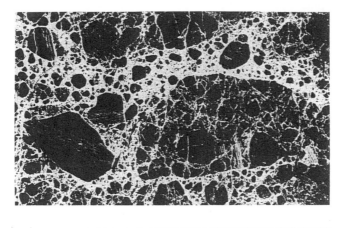

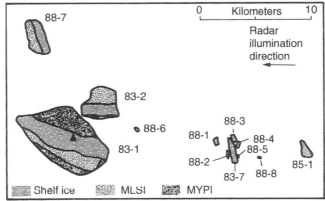

FIGURE 9.6 (Above) X-band HH SAR image of ice islands and pack ice near Ellef Ringnes Island, Canada, 19 February 1988. (Below) Interpretation of the ice islands. MLSI is multiyear landfast sea ice that was attached to the ice shelf at the time of calving; MYPI is multiyear pack ice that has subsequently accreted onto the ice island. Ice island "83-1" is also known as "Hobson's Choice." The bright feature located by a triangle is a research station on the ice island. (Reproduced from Jeffries and Sackinger (1990). Copyright American Geophysical Union, reproduced with permission of the American Geophysical Union.)

goal of an "on-demand" observational system, needed to provide tactical data for activities such as ship routing and operation of offshore hydrocarbon facilities, is still not fully met by such systems (Rossiter et al., 1995).

9.3 ICEBERG THICKNESS

Iceberg thickness is most commonly estimated from airborne or spaceborne VIR imagery, although impulse radar is also used (Rossiter and Gustajtis

1978). Analysis of VIR imagery yields the ice freeboard, i.e., the height above the water, from which the thickness can be calculated. The usual approach is through stereophotogrammetry (Løvås, Spring, and Holm 1993; Vefsnmo et al. 1989) to construct digital elevation models of icebergs. In a typical application, a mapping camera is flown at a height of 600 to 900 m, and the stereo-photographs can be used to construct a DEM with a height accuracy of the order of 10 cm. In the case of manual processing, Vefsnmo et al. (1989) found that the processing time could be as long as 3 hours per iceberg, although they also developed a simplified stereoscopic analysis with lower accuracy but a much faster processing time (10 minutes per iceberg). Computer-based stereo matching offers even greater savings of time.

A much simpler procedure for estimating iceberg thickness from aerial photography is described by Vefsnmo et al. (1989). It can be applied when the iceberg casts a shadow on the surrounding sea ice. The iceberg freeboard h is given by

$$h = l \tan \alpha + h_0 \tag{9.1}$$

where l is the length of the shadow, α is the solar elevation, and h_0 is the freeboard of the sea ice. This method is, however, only suitable for icebergs that are reasonably tabular in form.

10 Conclusions

10.1 WHAT HAS REMOTE SENSING REVEALED ABOUT THE CRYOSPHERE?

Spaceborne remote sensing of the cryosphere can be dated to roughly 1970. Since that time, the number, diversity, and quality of spaceborne instruments have increased, and progressively longer runs of self-consistent data have been developed. What have we learnt from this stream of information? Firstly, it is worth reiterating the point that spaceborne remote sensing permits some important measurements to be made that could not be performed in any other way. This is clearly true of synoptic-scale observations over time, so the fact that we now have a good understanding of the global distribution, interannual variation, and long-term trends of snow cover, sea ice, and the great ice sheets is a direct consequence of continuous monitoring from space. These results were presented and discussed in Sections 5.2, 6.2, and 8.3, respectively. Mapping ability in general has also increased dramatically since Swithinbank (1988) remarked that the far side of the Moon was better mapped than parts of Antarctica. Similarly, very little was known before 1978 (the year of the Seasat mission) about the topography of the Greenland ice sheet. Here again, spaceborne remote sensing has the obvious merits of rapid data collection and the fact that it is not necessary to deploy people on the ground in a hostile and logistically expensive environment.

The advantages of remote sensing are not confined to the largest-scale questions, such as those related to the ice sheets, the sea ice cover as a whole, or the snow cover of the northern hemisphere. Remote sensing has provided a set of tools, growing in range and sophistication over the years, for measuring and monitoring processes and phenomena on the Earth's surface. These tools have been applied to a wide range of phenomena relating to snow and ice, so that, for example, fluctuations in the margin of glaciers, glacier mass-balance, the dynamics of sea ice, inventories of icebergs, and the dates of freeze-up and thawing of river and lake ice have all been investigated on local and regional scales. This list is far from exhaustive. However, these may all be said to be to some extent expected consequences of the availability of remote sensing technologies. The application of remote sensing to the study of snow and ice has also produced some surprises, and these can be divided into two classes. Firstly there are the "geophysical" surprises, where new phenomena, or phenomena more extensive than expected, have been observed. Examples of this kind of surprise include the discovery of the "snow megadunes" in Antarctica and the observation of the rapid and extensive collapse of the Larsen ice shelf in Antarctica. The second class of surprise is the discovery that a remote sensing technique has the ability to detect a known phenomenon that it was not previously thought possible to distinguish. Examples of this kind of surprise include the ability of radio echo-sounding and ground-penetrating radar to reveal internal layering and other subglacial features such as lakes in glaciers, the clues to internal structure and the flow regime of glaciers and ice sheets from detailed surface topography, the ability of interferometric SAR to monitor the tidal flexure of ice shelves, and the ability of SAR imagery to indicate whether or not a body of freshwater ice is frozen to its bed.

10.2 WHAT ARE THE REMAINING TECHNICAL CHALLENGES?

10.2.1 GENERAL DIFFICULTIES

Although our ability to measure and understand the cryosphere has increased spectacularly since 1970, a number of challenges still remain. One of the most pervasive of these is cloud cover. As was noted in Chapter 2, and throughout the discussion of remote sensing of specific aspects of snow and ice presented in Chapters 5 to 9, cloud is generally opaque to visible, near-infrared and thermal infrared radiation, and while major advances have been made in the application of microwave techniques, these shorter-wavelength methods cannot be dispensed with. Figure 10.1 illustrates the global mean cloud amount for the four quarters of the year, and shows that in the summer months of the Arctic and sub-Arctic this is generally above 50% over land and 75% over parts of the ocean. While most of Antarctica experiences a mean cloud cover of less than 50% during the austral summer, the cloud cover over the Southern Ocean

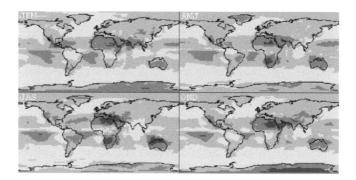

FIGURE 10.1 Average cloud amount for the four quarters of the year, calculated from data supplied by the International Satellite Cloud Climatology Program ISCCP (http://isccp.giss.nasa.gov/products/onlineData.html). Progressively lighter shades of gray represent average cloud amounts of 0 to 25%, 25 to 50%, 50 to 75%, and 75 to 100%, respectively.

during these months is mostly above 75%. The high prevalence of cloud cover is a particular problem for the narrower-swath sensors, which give poorer temporal resolution. For example, if a satellite-borne sensor only provides an opportunity to observe a location of interest once a fortnight, and if there is an 80% probability of cloud cover on each such opportunity, the average time interval between cloud-free observations will be 10 weeks. This interval may be so long as to deny the possibility of obtaining an image sufficiently close to a desired time such as the end of the melt season.

A related problem is the recognition of the presence of a partial cloud cover in an image. The only potential source of confusion is with snow cover, and as was noted in Sections 5.2.1 and 8.2, this is largely avoided if the instrument possesses a spectral band at around 1.6 μm where the reflectance of cloud is high and snow is low. Optically thin cirrus cloud remains difficult to discriminate from snow, and older VIR imagery may lack a spectral response at 1.6 μm so that cloud–snow discrimination can be difficult to perform consistently in long runs of imagery. For example, the Landsat satellites only acquired a 1.6 μm band with the advent of the Thematic Mapper instrument (1984 onward), while the AVHRR instruments have possessed such a band only since 1998.

Considering now the microwave systems, a number of general difficulties, not specific to particular aspects of the cryosphere, can also be identified. Spaceborne passive microwave systems are limited by diffraction to rather coarse spatial resolutions of typically tens of kilometers. This makes them unfit for investigations of small-scale phenomena — most of the world's glaciers and icebergs are not resolvable by satellite-based passive microwave radiometry. This is also a difficulty, though not so extreme, for radar altimetry. Although synthetic aperture radar and microwave scatterometry offer much higher spatial resolutions, most spaceborne systems to date have recorded a single variable — the backscattering coefficient at a single frequency and in a single polarization state. As has been noted several times (e.g., in the case of snow mapping, discussed in Section 5.2.1.2), this can introduce ambiguities into interpretation and classification. SAR interferometry remains a difficult technique to put into practice, because it is still something of a matter of luck whether a pair of observations separated by a few days will in fact have a suitable baseline between them and whether there will be adequate coherence between the observations to form an interferogram. The data processing for InSAR is also still largely in the domain of the specialist.

The laws of orbital dynamics impose some limitations on the possibilities of spaceborne observations. Most remote sensing satellites (other than those in geostationary orbits) are placed in *Sun-synchronous orbits*, which provides the valuable benefit of standardizing the time of day (and hence illumination conditions) at which images are acquired, but at the cost of limiting the latitudinal range of the orbit to between about 82° north and south. Since this brings the subsatellite track only to within about 900 km from the poles, it means that the imagery will not be truly global unless the swath width is at least 1800 km. Furthermore, different applications of spaceborne remotely sensed data can make incompatible demands on the design of the satellite orbit. Some applications require a spatially dense pattern of subsatellite tracks on the Earth's surface. An example of this is the case of a sensor with a very narrow swath width, such as a radar altimeter or laser profiler, where a dense orbital coverage is the only way of achieving a high spatial resolution. However, dense patterns of tracks imply that the time between successive revisits of a given location will be correspondingly long, so that the imagery may be unsuitable for studying dynamic phenomena, such as the motion of sea ice. To give a specific example, a satellite in a Sun-synchronous, nearly polar orbit takes approximately 100 minutes to orbit the Earth. If the orbital parameters are chosen in such a way that the subsatellite track repeats itself every 3 days, there will be roughly (3 days)/(100 minutes) ≈ 43 distinct orbital tracks, with a spacing of around 400 km at a latitude of 65°. To reduce this spacing to 10 km would require a repeat period 40 times as long, or about 4 months. Of course, this problem can be avoided (at greater expense) by using different satellites for different applications. Its severity is also decreased by increasing the swath width of the sensor, although this will generally be at the expense of a poorer spatial resolution. Finally, we should also note the difficulty of georeferencing satellite imagery from remote and largely featureless areas, discussed in Sections 3.2, 6.6, and 8.8, although this situation is continually improving as space agencies enhance the accuracy with which the position of a satellite, and the direction of view of a sensor, can be determined.

10.2.2 SPECIFIC DIFFICULTIES

Having considered the technical difficulties common to spaceborne observations of all the cryospheric phenomena discussed in this book, we now summarize the current limitations that are specific to each component individually. These are discussed in greater detail in the corresponding chapters.

The identification of *snow cover* is generally straightforward, although the presence of a forest cover (and much of the world's temporary and seasonal snow cover occurs in forested areas) or complex landscape greatly complicates the task. Techniques based on the analysis of passive microwave data are also difficult to apply in the presence of precipitation, or when the snow cover is thin. Estimates of *snow water equivalent* can currently be reliably made only using passive microwave data, with the usual limitations of spatial resolution, and only for dry snow. The effects of metamorphism and variation in grain size significantly complicate the task of estimating the SWE. Snow *depth* is particularly difficult to estimate from space.

The spatial extent of *sea ice* is generally reasonably accurately determined from space, and passive microwave methods are particularly valuable in this regard. However, measurement of ice *concentration* is less reliable. There is as yet no single universally accepted algorithm for ice concentration, and all current algorithms perform poorly during the summer, and in general whenever the surface is wet. Sea ice *thickness* is exceptionally difficult to measure from space.

In contrast with sea ice, remote sensing of *freshwater ice* is less well developed. One reason for this is undoubtedly the finer spatial scale of the phenomenon, which means that the ability to provide adequate spatial resolution of many lakes and rivers imposes more stringent demands on the spatial resolution of observing systems. As has already been noted, this tends to decrease the temporal resolution. Questions of spatial resolution aside, freshwater ice can be studied using both VIR and SAR imagery, although the detection of *black ice* remains a problem and no satisfactory technique currently exists for the determination of ice *thickness* from space.

Spaceborne remote sensing of *glaciers* and larger terrestrial ice masses is generally well developed. However, the identification of *surface facies* and of the *equilibrium line* from radar images remains somewhat ambiguous, as does the determination of *accumulation rates* from passive microwave data. Glacier *thickness* cannot currently be measured directly from space (although airborne and surface methods are valuable), but changes in ice volume can be monitored by measuring changes in *surface topography*. However, topographic measurements from space currently provide rather coarse spatial resolution, since they are limited by the orbital pattern of satellites and (in the case of radar altimetry) the spatial resolution of the instrument itself. For this reason, such topographic measurements are suitable for ice caps and ice sheets but are of limited use for smaller glaciers.

Icebergs are relatively straightforward to detect against a background of open water in spaceborne imagery, certainly in the case of the large tabular icebergs but also generally for smaller bergs. However, the majority of icebergs are less than 100 m wide, and these require observing systems with high spatial resolution. For the general reasons discussed in Section 10.2.1, this precludes synoptic monitoring and makes tracking of the *motion* of small icebergs difficult. The distinguishing of icebergs against a background of sea ice is much more challenging, and again, high-resolution imagery is required. Iceberg *thickness* cannot be determined directly from space, except perhaps in the case of large tabular bergs.

10.3   RECENT TRENDS AND SCOPE FOR FUTURE DEVELOPMENT

Notwithstanding the discussion of the shortcomings of spaceborne remote sensing in the previous section, a number of developments can be identified as showing promise for the future development of cryospheric monitoring. These include dedicated glaciological missions, or phases of missions, such as the Radarsat Antarctic Mapping Mission, IceSAT, and Cryosat, and the development of more flexible instruments that include combinations of observing parameters more suitable for snow and ice phenomena. Examples of this flexibility include radar altimeters that are less prone to the loss of tracking caused by abrupt changes in topographic height (Section 8.3), and synthetic aperture radars with a diversity of combinations of spatial resolution and swath width (Section 6.2). New applications of data from existing sensors are also emerging. For example, radar altimeter data show promise for the measurement of snow depth (Section 5.3.2), sea ice concentration (Section 6.2), and thickness (Section 6.5), while microwave scatterometry is also receiving increased attention, e.g., for sea ice concentration (Section 6.2). In perhaps a slightly different category can be placed observing techniques that have been available for some years but have not yet been widely exploited. This category includes SAR interferometry, where some applications such as glacier motion and topography are reasonably well developed, while others such as the ability to measure snow water equivalent (Section 5.3.1) and to identify the snow line on glaciers (Section 8.7) are comparatively poorly explored. The category also includes laser profiling, where major technological advances have been made for airborne systems in recent years (and from which particularly exciting results in glacier mapping are beginning to emerge, as discussed in Section 8.3) and where useful spaceborne data are becoming available. Perhaps the most eagerly awaited type of new spaceborne observing system is the multi-frequency, multipolarimetric SAR. This is likely to resolve many of the ambiguities inherent in single-parameter SAR images for a wide range of cryospheric applications, including snow mapping (Section 5.2.1.2), snow depth (Section 5.3.1), sea ice type (Section 6.3), and glacier facies mapping (Section 8.7).

Another promising area of development is the exploitation of data synergies. A classic example of data synergy is the well-established snow runoff model (Section 5.4), in which data from a variety of sources are combined in a Geographic Information System. Simple and also well-established examples include the use of land/sea masks for determining the extent and concentration of sea ice from VIR or passive microwave data, and the use of multiple data sources for global snow mapping (Section 5.2.3). Other examples include the combination of SAR or passive microwave data and land cover classification (e.g., from VIR data) for snow mapping (Sections 5.2.1.2, 5.2.3, and 5.3.2), of passive microwave and thermal infrared data to resolve ambiguities in the determination of snow extent (Sections 5.2.3 and 5.3.2), of scatterometer and passive microwave data for sea ice type (Section 6.3), of photoclinometry and radar altimeter data for enhancing the spatial resolution of a digital elevation model of an ice sheet (Section 8.3), and of passive microwave and radar altimeter data for estimating the accumulation rate over an ice sheet (Section 8.6). A particular class of data synergy is represented by the use of digital elevation model (DEM) data to augment the interpretation of imagery over land surfaces. The increasing availability of accurate, high-resolution DEMs from InSAR or laser profiling amplifies this trend. Examples of the use of DEMs include modeling shadows in VIR imagery (Section 5.2.1.1), correction of incidence-angle and viewing geometry effects (including terrain distortion) in VIR and SAR imagery (Sections 2.9.2, 5.2.1.2, and 5.5.2), and the separation of velocity and topographic components in InSAR analysis (Section 8.8).

Advances in methods of image classification and analysis also extend the promise of greater ability to extract useful data from remotely sensed imagery. Although the subject is evolving on many fronts, three areas in particular might be mentioned. The first of these is multitemporal data analysis, where variation over time increases the dimensionality of the dataset. An example of this was given in Section 6.3. The second area is texture analysis, which has not yet been as fully exploited in cryospheric applications as it might be. Lastly, neural network and expert system approaches show promise of substantially improving our ability to extract quantitative environmental information from remotely sensed data.

Bibliography

Abdalati, W., and K. Steffen. 1995. Passive microwave-derived snow melt regions on the Greenland ice sheet. *Geophysical Research Letters* 22 (7):787–790.

——. 1998. Accumulation and hoar frost effects on microwave emission on the Greenland ice-sheet dry-snow zones. *Journal of Glaciology* 44 (148):523–531.

Ackerman, S.A., K.I. Strabala, P.W.P. Menzel, R.A. Frey, C.C. Moeller, and L.E. Gumley. 1998. Discriminating clear sky from clouds with MODIS. *Journal of Geophysical Research* 103 (D24):32141–32157.

Adalgeirsdóttir, G., K. Echelmeyer, and W.D. Harrison. 1998. Elevation and volume changes on the Harding Icefield, Alaska. *Journal of Glaciology* 44 (148):570–582.

Ahlnäs, K., C.S. Lingle, W.D. Harrison, T.A. Heinrichs, and K.A. Echelmeyer. 1992. Identification of late-summer snow lines on glaciers in Alaska and the Yukon Territory with ERS-1 SAR imagery (abstract). *EOS Transactions* 73 (43):204.

Albertz, J., and K. Zelianeos. 1990. Enhancement of satellite image data by data cumulation. *Journal of Photogrammetry and Remote Sensing* 45 (3):161–174.

Alley, R.B. 1995. Resolved: the Arctic controls global climate change. In *Arctic Oceanography: Marginal Ice Zones and Continental Shelves*, edited by W.O. Smith and J.M. Grebmeier. Washington DC: American Geophysical Union.

Alley, R.B., C.A. Shuman, D.A. Meese, A.J. Gow, K.C. Taylor, K.M. Cuffey, J.J. Fitzpatrick, G. Spinelli, G.A. Zielinski, M. Ram, P.M. Grootes, and B. Elder. 1997. Visualising stratigraphic dating of the Greenland Ice Sheet Project 2 (GISP2) ice core: Basis, reproducibility, and application. *Journal of Geophysical Research* 102:26367–26381.

Andersen, T. 1982. Operational snow mapping by satellites. Paper read at Hydrological Aspects of Alpine and High Mountain Areas, at Exeter, U.K.

Andreassen, L.M., H. Elvehøy, and B. Kjøllmoen. 2002. Using aerial photography to study glacier changes in Norway. *Annals of Glaciology* 34:343–348.

Aniya, M., H. Sato, R. Naruse, P. Skvarca, and G. Casassa. 1996. The use of satellite and airborne imagery to inventory outlet glaciers of the southern Patagonia icefield, South America. *Photogrammetric Engineering and Remote Sensing* 62 (12):1361–1369.

Annan, A.P., and J.L. Davis. 1977. Impulse radar applied to ice thickness measurements and freshwater bathymetry. *Geological Society of Canada papers* 77–1B:63–65.

Archer, D.R., J.O. Bailey, E.C. Barrett, and D. Greenhill. 1994. The potential of satellite remote-sensing of snow over Great Britain in relation to cloud cover. *Nordic Hydrology* 25 (1–2):39–52.

Arcone, S.A., and A.J. Delaney. 1987. Airborne river-ice thickness profiling with helicopter-borne UHF short-pulse radar. *Journal of Glaciology* 33 (115):330–340.

Arcone, S.A., A.J. Delaney, and D.J. Calkins. 1989. Water detection in the coastal plains of the Arctic National Wildlife Refuge using helicopter-borne short-pulse radar. Hanover, NH: U.S. Army Cold Regions Research and Engineering Laboratory.

(Argentina), Servicio Meteorológico Nacional. 1992. Monitoring of a drifting iceberg in the South Atlantic. *Marine Observer* 62 (317):130–134.

Armstrong, R.L. 1985. Metamorphism in a subfreezing, seasonal snow cover: the role of thermal and vapor pressure conditions. PhD, University of Colorado, Boulder.

Armstrong, R.L., and M.J. Brodzik. 2002. Hemispheric-scale comparison and evaluation of passive-microwave snow algorithms. *Annals of Glaciology* 34:38–44.

Armstrong, R.L., A. Chang, A. Rango, and E. Josberger. 1993. Snow depths and grain-size relationships with relevance for passive microwave studies. *Annals of Glaciology* 17:171–176.

Avery, T.E., and G.L. Berlin. 1992. *Fundamentals of Remote Sensing and Airphoto Interpretation,* 5th ed. New York: Macmillan Publishing Company.

Baghdadi, N., J-P. Fortin, and M. Bernier. 1999. Accuracy of wet snow mapping using simulated Radarsat backscattering coefficients from observed snow cover characteristics. *International Journal of Remote Sensing* 20 (10):2049–2068.

Baghdadi, N., Y. Gauthier, and M. Bernier. 1997. Capability of multitemporal ERS-1 SAR data for wet-snow mapping. *Remote Sensing of Environment* 60 (2):174–186.

Baghdadi, N., C.E. Livingstone, and M. Bernier. 1998. Airborne C-band SAR measurements of wet snow-covered areas. *IEEE Transactions on Geoscience and Remote Sensing* 36 (6):1977–1981.

Bamber, J.L. 1994. Ice sheet altimeter processing scheme. *International Journal of Remote Sensing* 15 (4):925–938.

Bamber, J.L., and C.R. Bentley. 1994. A comparison of satellite altimetry and ice-thickness measurements of the Ross Ice Shelf, Antarctica. *Annals of Glaciology* 20:357–364.

Bamber, J.L., and R.A. Bindschadler. 1997. An improved elevation dataset for climate and ice-sheet modelling: validation with satellite imagery. *Annals of Glaciology* 25:439–444.

Bamber, J.L., and J.A. Dowdeswell. 1990. Remote-sensing studies of Kvitoyjokulen, an ice cap on Kvitoya, north-east Svalbard. *Journal of Glaciology* 36 (122):75–81.

Bamber, J.L., S. Ekholm, and W. Krabill. 1998. The accuracy of satellite radar altimeter data over the Greenland ice sheet. *Geophysical Research Letters* 25 (16):3177–3180.

Bamber, J.L., and A.R. Harris. 1994. The atmospheric correction for satellite infrared radiometer data in polar regions. *Geophysical Research Letters* 21 (19):2111–2114.

Bamber, J.L., R.L. Layberry, and S.P. Gogineni. 2001a. A new ice thickness and bed data set for the Greenland ice sheet. 1. Measurement, data reduction, and errors. *Journal of Geophysical Research* 106 (D24):33773–33780.

——. 2001b. A new ice thickness and bed data set for the Greenland ice sheet. 2. Relationship between dynamics and basal topography. *Journal of Geophysical Research* 106 (D24):33781–33788.

Baral, D.J., and R.P. Gupta. 1997. Integration of satellite sensor data with DEM for the study of snow cover distribution and depletion pattern. *International Journal of Remote Sensing* 18 (18):3889–3894.

Barber, D.G., and E.F. LeDrew. 1994. On the links between microwave and solar wavelength interactions with snow-covered first-year sea ice. *Arctic* 47:298–309.

Barber, D.G., T.N. Papakyriakou, and E.F. LeDrew. 1994. On the relationship between energy fluxes, dielectric properties, and microwave scattering over snow covered first-year sea ice during the spring transition period. *Journal of Geophysical Research* 99:22401–22411.

Barber, D.G., T.N. Papakyriakou, E.F. LeDrew, and M.E. Shokr. 1995. An examination of the relation between the spring period evolution of the scattering coefficient σ^0 and radiative fluxes over landfast sea-ice. *International Journal of Remote Sensing* 16:3343–3363.

Bardel, P., A.G. Fountain, D.K. Hall, and R. Kwok. 2002. Synthetic aperture radar detection of the snowline on Commonwealth and Howard Glaciers, Taylor Valley, Antarctica. *Annals of Glaciology* 34:177–183.

Barnett, T.P., L. Dümenil, U. Schlese, E. Roeckner, and M. Latif. 1989. The effect of Eurasian snow cover on regional and global climatic variations. *Journal of Atmospheric Science* 46 (5):661–685.

Barton, I.J., A.M. Zavody, D.M. O'Brien, D.R. Cutten, R.W. Saunders, and D.T. Llewellyn-Jones. 1989. Theoretical algorithms for satellite-derived sea surface temperatures. *Journal of Geophysical Research* 94 (D3):3365–3375.

Basist, A., and N.C. Grody. 1994. Identification of snowcover, using SSM/I measurements. Paper read at American Meteorological Society: Sixth Conference on Climatic Variations, at Nashville.

Bauer, P., and N. Grody. 1995. The potential of combining SSM/I and SSMT/2 measurements to improve the identification of snow cover and precipitation. *IEEE Transactions on Geoscience and Remote Sensing* 33:252–261.

Baumgartner, M.F., and G. Apfl. 1993. Alpine snow cover analysis system. Paper read at Sixth AVHRR Data Users' Meeting, at Belgirate.

Baumgartner, M.F., G. Apfl, and T. Holzer. 1994. Monitoring Alpine snow cover variations using NOAA-AVHRR data. Paper read at International Geoscience and Remote Sensing Symposium. Surface and Atmospheric Remote Sensing: Technologies, Data Analysis and Interpretation, at Pasadena.

Baxter, J.P. 1991. Soviet satellite imagery: the photographic alternative to digital remote sensing data. *Mapping Awareness* 5:30–33.

Bayr, K., D.K. Hall, and W.M. Kovalick. 1994. Observations on glaciers in the eastern Austrian Alps using satellite data. *International Journal of Remote Sensing* 15 (9):1733–1742.

Beaven, S.G., S.P. Gogineni, S. Tjuatja, and A.K. Fung. 1997. Model-based interpretation of ERS-1 SAR images of Arctic sea ice. *International Journal of Remote Sensing* 18 (12):2483–2503.

Beltaos, S., D.J. Calkins, L.W. Gatto, T.D. Prowse, S. Reedyk, G.J. Scrimgeour, and S.P. Wilkins. 1993. Physical effects of river ice. In *Environmental Aspects of River Ice*, edited by T.D. Prowse and N.C. Gridley. Saskatoon: National Hydrology Research Institute, Canada.

Beltrami, H., and A.E. Taylor. 1994. Records of climatic change in the Canadian Arctic: combination of geothermal and oxygen isotope data yields high resolution ground temperature histories. *EOS Transactions* 75 (44):75.

Benson, C.S. 1961. Stratigraphic studies in the snow and firn of the Greenland Ice Sheet. *Folia Geographica Danica* 9:13–37.

Bindschadler, R. 1998. Monitoring ice sheet behavior from space. *Reviews of Geophysics* 36 (1):79–104.

Bindschadler, R., M. Fahnestock, and R. Kwok. 1992. Monitoring of the Greenland ice sheet using ERS-1 synthetic aperture radar imagery. Paper read at Space at the Service of our Environment: First ERS-1 Symposium, at Cannes.

Bindschadler, R., T.A. Scambos, H. Rott, P. Skvarca, and P. Vornberger. 2002. Ice dolines on Larsen Ice Shelf, Antarctica. *Annals of Glaciology* 34:283–290.

Bindschadler, R.A., K.C. Jezek, and J. Crawford. 1987. Glaciological investigations using the Synthetic Aperture Radar imaging system. *Annals of Glaciology* 9:11–19.

Bindschadler, R.A., and P.L. Vornberger. 1990. AVHRR imagery reveals Antarctic ice dynamics. *EOS Transactions* 71 (23):741–742.

———. 1992. Interpretation of SAR imagery of the Greenland ice sheet using coregistered TM imagery. *Remote Sensing of Environment* 42 (3):167–175.

———. 1994. Detailed elevation map of ice stream C using satellite imagery and airborne radar. *Annals of Glaciology* 20:327–335.

Bingham, A.W., and W.G. Rees. 1999. Construction of a high-resolution DEM of an Arctic ice cap using shape-from-shading. *International Journal of Remote Sensing* 20 (15–16):3231–3242.

Bishop, M.P., J.F. Shroder, B.L. Hickman, and L. Copland. 1998. Scale-dependent analysis of satellite imagery for characterization of glacier surfaces in the Karakoram Himalaya. *Geomorphology* 21 (3–4):217–232.

Bjørgo, E., O.M. Johannessen, and M.W. Miles. 1997. Analysis of merged SMMR-SSMI time series of Arctic and Antarctic sea ice parameters 1978–1995. *Geophysical Research Letters* 24 (4):413–416.

Bobylev, L.P., K.Y. Kondratyev, and O.M. Johannessen, eds. 2003. *Arctic Environment Variability in the Context of Global Change*. Chichester: Springer-Praxis.

Bochert, A. 1999. Airborne line scanner measurements for ERS-1 SAR interpretation of sea ice. *International Journal of Remote Sensing* 20 (2):329–348.

Bogorodsky, V.V., C.R. Bentley, and P. Gudmandsen. 1985. *Radioglaciology*. Dordrecht: Reidel.

Borodulin, V.V. 1989. Use of radar images from the Kosmos-1500 and -1766 satellites to describe ice conditions on inland bodies of water. [Ispol'zovaniye radiolokatsionnykh snimkov s ISZ "Kosmos-1500, 1766" dlya kharakteristiki ledovoy obstanovki na vnutrennikh vodoyemakh.] *Soviet Meteorology and Hydrology* 5:70–74.

Borodulin, V.V., and V.G. Prokacheva. 1983. Studying lake ice regimes by remote sensing methods. Paper read at Hydrological Applications of Remote Sensing and Remote Data Transmission, 1985, at Hamburg.

Bourdelles, B., and M. Fily. 1993. Snow grain-size determination from Landsat imagery over Terre Adélie, Antarctica. *Annals of Glaciology* 17:86–92.

Braun, M., F. Rau, H. Saurer, and H. Gossmann. 2000. Development of radar glacier zones on the King George Island ice cap, Antarctica, during austral summer 1996/97 as observed in ERS-2 SAR data. *Annals of Glaciology* 31:357–363.

Brenner, A.C., R.A. Bindschadler, R.H. Thomas, and H.J. Zwally. 1983. Slope-induced errors in radar altimetry over continental ice sheets. *Journal of Geophysical Research* 88:1617–1623.

Bromwich, D.H., T.R. Parish, and C.A. Zorman. 1990. The confluence zone of the intense katabatic winds at Terra Nova Bay, Antarctica, as derived from airborne sastrugi surveys and mesoscale numerical modeling. *Journal of Geophysical Research* 95 (D5):5495–5509.

Brown, I.A., M.P. Kirkbride, and R.A. Vaughan. 1999. Find the firn line! The suitability of ERS-1 and ERS-2 SAR data for the analysis of glacier facies on Icelandic icecaps. *International Journal of Remote Sensing* 20 (15):3217–3230.

Brown, J., O.J. Ferrians, J.A. Heginbottom, and E.S. Melnikov. 1998. *Circum-arctic map of permafrost and ground-ice conditions*. National Snow and Ice Data Center/World Data Center for Glaciology 1998 (cited 26 May 2004). Available from http://nsidc.org/data/ggd318.html.

Brown, R.D. 2000. Northern hemisphere snow cover variability and change. *Journal of Climate* 13:2339–2355.

Brown, R.D., and R.O. Braaten. 1998. Spatial and temporal variability of Canadian snow depths 1946–1995. *Atmosphere and Ocean* 36 (1):37–54.

Bryan, M.L., and R.W. Larson. 1975. The study of freshwater lake ice using multiplexed imaging radar. *Journal of Glaciology* 14 (72):445–457.

Budd, W.F. 1970. Ice flow over bedrock perturbations. *Journal of Glaciology* 9 (55):29–48.

Budyko, M.I. 1966. Polar ice and climate. Paper read at Symposium on the Arctic Heat Budget and Atmospheric Circulation, at Santa Monica.

Burakov, D.A., and others. 1996. A technique for determining snow cover of a river basin from satellite data for operational runoff forecasts. [Metodika opredeleniya zasnezhennosti rechnogo basseyna po sputnikovym dannym dlya operativnykh prognozov stoka.] *Russian Meteorology and Hydrology* 8:58–65.

Burns, B.A., D.J. Cavalieri, M.R. Keller, W.J. Campbell, T.C. Grenfell, G.A. Maykut, and P. Gloersen. 1987. Multisensor comparison of ice concentration estimates in the marginal ice zone. *Journal of Geophysical Research* 92 (C7):6843–6856.

Campbell, J.B. 1996. *Introduction to Remote Sensing*, 2nd ed. London: Taylor and Francis.

Campbell, W.J., W.F. Weeks, R.O. Ramseier, and P. Gloersen. 1975. Geophysical studies of floating ice by remote sensing. *Journal of Glaciology* 15 (73):305–328.

Campbell, W.J., P. Gloersen, E.G. Josberger, O.M. Johannessen, P.S. Guest, N. Mognard, R. Shuchman, B.A. Burns, N. Lannelongue, and K.L. Davidson. 1987. Variations of mesoscale and large-scale ice morphology in the 1984 marginal ice-zone experiment as observed by microwave remote-sensing. *Journal of Geophysical Research — Oceans* 92 (C7):6805-6824.

Carroll, S.S., and T.R. Carroll. 1989. Effect of uneven snow cover on airborne snow water equivalent estimates obtained by measuring terrestrial gamma radiation. *Water Resource Research* 25 (7):101–115.

Carsey, F.D., R.G. Barry, and W.F. Weeks. 1992. Introduction. In *Microwave Remote Sensing of Sea Ice*, edited by F.D. Carsey. Washington DC: American Geophysical Union.

Casassa, G., K. Smith, A. Rivera, J. Araos, M. Schnirich, and C. Schneider. 2002. Inventory of glaciers in isla Riesco, Patagonia, Chile, based on aerial photography and satellite imagery. *Annals of Glaciology* 34:373–378.

Casassa, G., and J. Turner. 1991. Dynamics of the Ross Ice Shelf. *Eos* 72 (44):473–481.

Cavalieri, D.J., J.P. Crawford, M.R. Drinkwater, D.T. Eppler, L.D. Farmer, R.R. Jentz, and C.C. Wackerman. 1991. Aircraft active and passive microwave validation of sea ice concentration from the Defense Meteorological Satellite Program Special Sensor Microwave Imager. *Journal of Geophysical Research* 96:21989–22008.

Cavalieri, D.J., P. Gloersen, and W.J. Campbell. 1984. Determination of sea ice parameters with the Nimbus-7 SMMR. *Journal of Geophysical Research* 89:5355–5369.

Cavalieri, D.J., C.L. Parkinson, P. Gloersen, J.C. Comiso, and H.J. Zwally. 1999. Deriving long-term time series of sea ice cover from satellite passive-microwave multisensor data sets. *Journal of Geophysical Research* 104 (C7):15803–15814.

Chacho, E.F., S.A. Arcone, and A.J. Delaney. 1992. Location and detection of winter water supplies on the north slope of Alaska. Paper read at 43rd Arctic Science Conference, at Valdez, Alaska.

Chang, A.T.C., and L.S. Chiu. 1991. Satellite estimation of snow water equivalent: classification of physiographic regimes. Paper read at IGARSS. '91: International Geoscience and Remote Sensing Symposium: Global Monitoring for Earth Management, at Espoo, Finland.

——. 1990. Satellite sensor estimates of Northern Hemisphere snow volume. *International Journal of Remote Sensing* 11 (1):167–171.

Chang, A.T.C., J.L. Foster, and D.K. Hall. 1987. Nimbus-7 SMMR derived global snow cover parameters. *Annals of Glaciology* 9:39–44.

Chang, A.T.C., J.L. Foster, D.K. Hall, A. Rango, and B.K. Hartline. 1981. Snow water equivalent determination by microwave radiometry. In *NASA Technical Memoranda.* Greenfield, Maryland: NASA Goddard Space Flight Center.

Chang, A.T.C., and L. Tsang. 1992. A neural network approach to inversion of snow water equivalent from passive microwave measurements. *Nordic Hydrology* 23 (3):173–182.

Chang, T.C., P. Gloersen, T. Schmugge, T.T. Wilheit, and H.J. Zwally. 1976. Microwave emission from snow and glacier ice. *Journal of Glaciology* 16 (74):23–39.

Chase, J.R., and R.J. Holyer. 1990. Estimation of sea ice type and concentration by linear unmixing of Geosat altimeter waveforms. *Journal of Geophysical Research* 95 (C10):18015–18025.

Choi, E.M. 1999. The Radarsat Antarctic mapping mission. *IEEE Aerospace Electronics Systems Magazine* 14 (5):3–5.

Choudhury, B.J., and A.T.C. Chang. 1979. Two-stream theory of reflectance of snow. *IEEE Transactions on Geoscience Electronics* GE17:63–68.

Christensen, E.L., N. Reeh, R. Forsberg, J.H. Jörgensen, N. Skou, and K. Woelders. 2000. A low-cost glacier-mapping system. *Journal of Glaciology* 46 (154):531–537.

Cline, D.W. 1993. Measuring alpine snow depths by digital photogrammetry. Part 1. Conjugate point identification. Paper read at Eastern Snow Conference, at Quebec.

Cline, D.W., R.C. Bales, and J. Dozier. 1998. Estimating the spatial distribution of snow in mountain basins using remote sensing and energy balance modeling. *Water Resources Research* 34 (5):1275–1285.

Colby, J.D. 1991. Topographic normalisation in rugged terrain. *Photogrammetric Engineering and Remote Sensing* 57 (5):531–537.

Collins, M.J., and W.J. Emery. 1988. A computational method for estimating sea ice motion in sequential Seasat synthetic aperture radar imagery by matched filtering. *Journal of Geophysical Research* 93 (C8):9241–9251.

Collins, M.J., C.E. Livingstone, and R.K. Raney. 1997. Discrimination of sea ice in the Labrador marginal ice zone from synthetic aperture radar image texture. *International Journal of Remote Sensing* 18 (3):535–571.

Comiso, J. 1990. Arctic multiyear ice classification and summer ice cover using passive microwave satellite data. *Journal of Geophysical Research* 95:13411–13422.

———. 2002. Correlation and trend studies of the sea-ice cover and surface temperatures in the Arctic. *Annals of Glaciology* 34:420–428.

Comiso, J.C. 1983. Sea ice effective microwave emissivities from satellite passive microwave and infrared observations. *Journal of Geophysical Research* 88 (C12):7686–7704.

———. 2000. Variability and trends in Antarctic surface temperatures from in situ and satellite infrared measurements. *Journal of Climate* 13 (10):1674–1696.

Comiso, J.C., D.J. Cavalieri, C.L. Parkinson, and P. Gloersen. 1997. Passive microwave algorithms for sea ice concentration: a comparison of two techniques. *Remote Sensing of Environment* 60 (3):357–384.

Comiso, J.C., P. Wadhams, W.B. Krabill, R.N. Swift, J.P. Crawford, and W.B. Tucker. 1991. Top bottom multisensor remote-sensing of Arctic sea ice. *Journal of Geophysical Research — Oceans* 96 (C2):2693–2709.

Cooper, D.W., R.A. Mueller, and R.J. Schertler. 1976. Remote profiling of lake ice using an S-band short-pulse radar aboard an all-terrain vehicle. *Radio Science* 11:375–381.

Cracknell, A.P., and L.W.B. Hayes. 1991. *Introduction to Remote Sensing*. London: Taylor and Francis.

Cumming, W. 1952. The dielectric properties of ice and snow at 3.2 cm. *Journal of Applied Physics* 23:768–773.

Dahl, J.B., and H. Ødegaard. 1970. Areal measurement of water equivalents of snow deposits by means of natural radioactivity in the ground. Paper read at Symposium on Isotope Hydrology, at Vienna.

Dammert, P.B.G., M. Leppäranta, and J. Askne. 1998. SAR interferometry over Baltic Sea ice. *International Journal of Remote Sensing* 19 (16):3019–3037.

Das, S.B., R.B. Alley, D.B. Reusch, and C.A. Shuman. 2002. Temperature variability at Siple Dome, West Antarctica, derived from ECMWF re-analyses, SSM/I and SMMR brightness temperatures and AWS records. *Annals of Glaciology* 34:106–112.

Dash, M.K., S.M. Bhandari, N.K. Vyas, N. Khare, A. Mitra, and P.C. Pandey. 2001. Oceansat-MSMR imaging of the Antarctic and the Southern Polar Ocean. *International Journal of Remote Sensing* 22 (16):3253–3259.

Datcu, M. 1997. A new image formation model for the segmentation of the snow cover in mountainous areas. Paper read at EARSeL Workshop on Remote Sensing of Land Ice and Snow, at Freiburg.

Davis, C.H. 1993. A surface and volume retracking algorithm for ice sheet satellite altimetry. *IEEE Transactions on Geoscience and Remote Sensing* 31:811–818.

———. 1995. Synthesis of passive microwave and radar altimeter data for estimating accumulation rates of dry polar snow. *International Journal of Remote Sensing* 16:2055–2067.

Davis, D.T., Z.X. Chen, J.N. Hwang, A.T.C. Chang, and L. Tsang. 1993. Retrieval of snow parameters by iterative inversion of a neural-network. *IEEE Transactions on Geoscience and Remote Sensing* 31 (4):842–852.

De Sève, D., M. Bernier, J-P. Fortin, and A. Walker. 1997. Preliminary analysis of the snow microwave radiometry using the SSM/I passive microwave data: the case of La Grande River watershed (Quebec). *Annals of Glaciology* 25:353–361.

Demuth, M., and A. Pietroniro. 1999. Inferring glacier mass balance using RADARSAT: Results from Peyto Glacier, Canada. *Geografiska Annaler Series a — Physical Geography* 81A (4):521–540.

Derksen, C., A. Walker, E. LeDrew, and B. Goodison. 2002. Time-series analysis of passive-microwave-derived central North American snow water equivalent imagery. *Annals of Glaciology* 34:1–7.

Deser, C., J.E. Walsh, and M.S. Timlin. 2000. Arctic sea ice variability in the context of recent atmospheric circulation trends. *Journal of Climate* 13:617–633.

Dickson, R.R., T.J. Osborn, J.W. Hurrell, J. Meincke, J. Blindheim, B. Adlandsvik, and others. 2000. The Arctic Ocean response to the North Atlantic Oscillation. *Journal of Climate* 13:2671–2696.

Doake, C.S.M., H.F.J. Corr, H. Rott, P. Skvarca, and N.W. Young. 1998. Breakup and conditions for stability of the northern Larsen Ice Shelf, Antarctica. *Nature* 391 (6669):778–780.

Dobrowolski, A., and R. Gronet. 1990. Use of airborne remote sensing for assessment of intensity of slush ice transport in river. Paper read at IAHR 10th Symposium on Ice, at Espoo.

Dowdeswell, J.A. 1989. On the nature of the Svalbard icebergs. *Journal of Glaciology* 35:224–234.

Dowdeswell, J.A., A.F. Glazovsky, and Y.Y. Macheret. 1995. Ice divides and drainage basins on the ice caps of Franz Josef Land, Russian High Arctic, defined from Landsat, KFA-1000, and ERS-1 SAR satellite imagery. *Arctic and Alpine Research* 27 (3):264–270.

Dowdeswell, J.A., M.R. Gorman, A.F. Glazovsky, and Yu.Ya. Macheret. 1994. Evidence for floating ice shelves in Franz Josef Land, Russian High Arctic. *Arctic and Alpine Research* 26 (1):86–92.

Dowdeswell, J.A., M.R. Gorman, Yu.Ya. Macheret, M.Yu. Moskalevsky, and J.O. Hagen. 1993. Digital comparison of high resolution Sojuzkarta KFA-1000 imagery of ice masses with Landsat and SPOT data. *Annals of Glaciology* 17:105–112.

Dowdeswell, J.A., and N.F. McIntyre. 1986. The saturation of Landsat MSS detectors over large ice masses. *International Journal of Remote Sensing* 7 (1):151–164.

Dowdeswell, J.A., W.G. Rees, and A.D. Diament. 1993. ERS-1 SAR investigations of snow and ice facies in the European High Arctic. Paper read at Space at the Service of Our Environment: Second ERS-1 Symposium, at Hamburg.

Dowdeswell, J.A., and M. Williams. 1997. Surge-type glaciers in the Russian High Arctic identified from digital satellite imagery. *Journal of Glaciology* 43 (145):489–494.

Dozier, J. 1984. Snow reflectance from Landsat-4 Thematic Mapper. *IEEE Transactions on Geoscience and Remote Sensing* GE22 (3):323–328.

——. 1989. Spectral signature of Alpine snow cover from the Landsat Thematic Mapper. *Remote Sensing of Environment* 28:9–22.

Dozier, J., and J. Frew. 1990. Rapid calculation of terrain parameters for radiation modeling from digital elevation data. *IEEE Transactions on Geoscience and Remote Sensing* GE28 (5):963–969.

Dozier, J., and D. Marks. 1987. Snow mapping and classification from Landsat Thematic Mapper data. *Annals of Glaciology* 9:97–103.

Dozier, J., S.R. Schneider, and D.F. McGinnis. 1981. Effect of grain size and snowpack water equivalence on visible and near-infrared satellite observations of snow. *Water Resources Research* 17:1213–1221.

Drewry, D.J. 1981. Radio echo sounding of ice masses: principles and applications. In *Remote Sensing in Meterology, Oceanography and Hydrology*, edited by A.P. Cracknell. Chichester: Ellis Horwood.

——. 1983. Antarctic ice sheet thickness and volume. In *Antarctica: Glaciological and Geophysical Folio*, edited by D.J. Drewry. Cambridge: Cambridge University Press.

Drinkwater, M.R. 1991. Ku-band airborne radar altimeter observations of sea ice during the 1984 Marginal Ice Zone Experiment. *Journal of Geophysical Research* 96 (C3):4555–4572.

Drinkwater, M.R., R. Kwok, E. Rignot, H. Israelson, R.G. Onstott, and D.P. Winebrenner. 1992. Potential applications of polarimetry to the classification of sea ice. In *Microwave Remote Sensing of Sea Ice*, edited by F.D. Carsey. Washington DC: American Geophysical Union.

Drinkwater, M.R., R. Kwok, D.P. Winebrenner, and E. Rignot. 1991. Multifrequency polarimetric synthetic aperture radar observations of sea ice. *Journal of Geophysical Research — Oceans* 96 (C11):20679–20698.

Drinkwater, M.R., and V.A. Squire. 1989. C-band SAR observations of marginal ice zone rheology in the Labrador Sea. *IEEE Transactions on Geoscience and Remote Sensing* GE27:522–534.

Duguay, C.R., and E.F. LeDrew. 1992. Estimating surface reflectance and albedo over rugged terrain from Landsat-5 Thematic Mapper. *Photogrammetric Engineering and Remote Sensing* 58 (5):551–558.

Ebbesmeyer, C.C., A. Okubo, and H.J.M. Helset. 1980. Description of iceberg probability between Baffin Bay and the Grand Banks using a stochastic model. *Deep-Sea Research* 27A:975–986.

Echelmeyer, K.A., W.D. Harrison, C.F. Larsen, J. Sapiano, J.E. Mitchell, J. DeMallie, and B. Rabus. 1996. Airborne surface elevation measurements of glaciers: a case study in Alaska. *Journal of Glaciology* 42 (142):538–547.

Eisen, O., U. Nixdorf, F. Wilhelms, and H. Miller. 2002. Electromagnetic wave speed in polar ice: validation of the common-midpoint technique with high-resolution dielectric profiling and γ-density measurements. *Annals of Glaciology* 34:150–156.

Ekholm, S., R. Forsberg, and J.M. Brozena. 1995. Accuracy of satellite altimeter elevations over the Greenland ice sheet. *Journal of Geophysical Research* 100 (C2):2687–2696.

El Naggar, S., C. Garrity, and R.O. Ramseier. 1998. The modelling of sea ice melt-water ponds for the High Arctic using an Airborne line scan camera, and applied to the Satellite Special Sensor Microwave/Imager (SSM/I). *International Journal of Remote Sensing* 19 (12):2373–2394.

Elachi, C., M.L. Bryan, and W.F. Weeks. 1976. Imaging radar observations of frozen Arctic lakes. *Remote Sensing of Environment* 5:169–175.

Emery, W.J., C.W. Fowler, J. Hawkins, and R.H. Preller. 1991. Fram Strait satellite image-derived ice motions. *Journal of Geophysical Research* 96 (C5):8917–8920.

Engeset, R.V. 2000. Change detection and monitoring of glaciers and snow using satellite microwave imaging. PhD. University of Oslo, Oslo.

Engeset, R.V., J. Kohler, K. Melvold, and B. Lundén. 2002. Change detection and monitoring of glacier mass balance and facies using ERS SAR winter images over Svalbard. *International Journal of Remote Sensing* 23 (10):2023–2050.

Engeset, R.V., and R.S. Ødegård. 1999. Comparison of annual changes in winter ERS-1 SAR images and glacier mass balance of Slakbreen, Svalbard. *International Journal of Remote Sensing* 20 (2):259–271.

Engeset, R.V., and D.J. Weydahl. 1998. Analysis of glaciers and geomorphology on Svalbard using multitemporal ERS-1 SAR images. *IEEE Transactions on Geoscience and Remote Sensing* 36 (6):1879–1887.

Eppler, D.T., L.D. Farmer, A.W. Lohanick, M.R. Anderson, D.J. Cavalieri, J. Comiso, P. Gloersen, C. Garrity, T.C. Grenfell, M. Hallikainen, J.A. Maslanik, C. Mätzler, R.A. Melloh, I. Rubinstein, and C.T. Swift. 1992. Passive microwave signatures of sea ice. In *Microwave Remote Sensing of Sea Ice*, edited by F.D. Carsey. Washington DC: American Geophysical Union.

Eyton, R. 1989. Low-relief topographic enhancement in a Landsat snow-cover scene. *Remote Sensing of Environment* 27:105–118.

Fahnestock, M.A., W. Abdalati, and C.A. Shuman. 2002. Long melt seasons on ice shelves of the Antarctic Peninsula: an analysis using satellite-based microwave emission measurements. *Annals of Glaciology* 34:127–133.

Fahnestock, M.A., and R.A. Bindschadler. 1993. Description of a program for SAR investigation of the Greenland ice sheet and an example of margin detection using SAR. *Annals of Glaciology* 17:332–336.

Fahnestock, M.A., R.A. Bindschadler, R. Kwok, and K.C. Jezek. 1993. Greenland ice sheet surface properties and ice flow from ERS-1 SAR imagery. *Science* 262:1530–1534.

Fahnestock, M.A., T.A. Scambos, C.A. Shuman, R.J. Arthern, D.P. Winebrenner, and R. Kwok. 2000. Snow megadune fields on the East Antarctic Plateau: extreme atmosphere–ice interaction. *Geophysical Research Letters* 27 (22):3719–3722.

Fairbanks, R.G. 1989. A 17,000-year glacio-eustatic sea level record: influence of glacial melting rates on the Younger Dryas event and deep-ocean circulation. *Nature* 342 (6520):637–649.

Fastook, J.L., H.H. Brecher, and T.J. Hughes. 1995. Derived bedrock elevations, strain rates and stresses from measured surface elevations and velocities: Jakobshavns Isbrae, Greenland. *Journal of Glaciology* 41 (137):161–173.

Fatland, D.R., and C.S. Lingle. 1998. Analysis of the 1993–1995 Bering Glacier (Alaska) surge using differential SAR interferometry. *Journal of Glaciology* 44 (148):532–546.

Favey, E., A. Geiger, G.H. Gudmundsson, and A. Wehr. 1999. Evaluating the potential of an airborne laser scanning system for measuring volume changes of glaciers. *Geografiska Annaler Series a — Physical Geography* 81:555–561.

FENCO. 1987. Optimum deployment of TODs (TIROS Ocean Drifters) to derive ocean currents for iceberg drift forecasting. Ontario: Meteorological Services Branch, Atmospheric Environment Division.

Ferrigno, J.G., and W.G. Gould. 1987. Substantial changes in the coastline of Antarctic revealed by satellite imagery. *Polar Record* 23 (146):577–583.

Ferrigno, J.G., B.K. Lucchitta, K.F. Mullins, A.L. Allison, R.J. Allen, and W.G. Gould. 1993. Velocity measurements and changes in position of Thwaites

Glacier/iceberg tongue from aerial photography, Landsat images and NOAAA VHRR data. *Annals of Glaciology* 17:239–244.

Ferrigno, J.G., R.S. Williams, E. Rosanova, B.K. Lucchitta, and C. Swithinbank. 1998. Analysis of coastal change in Marie Byrd Land and Ellsworth Land, West Antarctica, using Landsat imagery. *Annals of Glaciology* 27:33–40.

Fetterer, F., and N. Untersteiner. 1998. Observations of melt ponds on Arctic sea ice. *Journal of Geophysical Research* 103 (C11):24821–24835.

Fetterer, F.M., M.R. Drinkwater, K.C. Jezek, S.W.C. Laxon, R.G. Onstott, and L.M.H Ulander. 1992. Sea ice altimetry. In *Microwave Remote Sensing of Sea Ice*, edited by F.D. Carsey. Washington DC: American Geophysical Union.

Filin, S., and B. Csathó. 2002. Improvement of elevation accuracy for mass-balance monitoring using in-flight laser calibration. *Annals of Glaciology* 34:330–334.

Fily, M., B. Bourdelles, J.P. Dedieu, and C. Sergent. 1997. Comparison of in situ and Landsat thematic mapper derived snow grain characteristics in the Alps. *Remote Sensing of Environment* 59 (3):452–460.

Fily, M., J.P. Dedieu, and Y. Durand. 1999. Comparison between the results of a snow metamorphism model and remote sensing derived snow parameters in the Alps. *Remote Sensing of Environment* 68 (3):254–263.

Fily, M., and D.A. Rothrock. 1991. Opening and closing of sea ice leads: digital measurements from synthetic aperture radar. *Journal of Geophysical Research* 95 (C1):789–796.

Flint, R.F. 1971. *Glacial and Quaternary Geology*. 892 vols. New York: John Wiley.

Forster, R.R., and B.L. Isacks. 1994. The Patagonian icefields revealed by space shuttle synthetic aperture radar (SIR-C/X-SAR) (abstract). *EOS Transactions* 75 (44):226.

Forster, R.R., B.L. Isacks, and S.B. Das. 1996. Shuttle imaging radar (SIR-C/X-SAR) reveals near-surface properties of the South Patagonian ice field. *Journal of Geophysical Research* 101 (E10):23169–23180.

Forster, R.R., K.C. Jezek, J. Bolzan, F. Baumgartner, and S.P. Gogineni. 1999. Relationships between radar backscatter and accumulation rates on the Greenland ice sheet. *International Journal of Remote Sensing* 20 (15):3131–3147.

Forsythe, K.W. 1999. Developing snowpack models in the Kalkhochalpen region. PhD. University of Salzburg, Salzburg.

Foster, J., D. Schultz, and W.C. Dallam. 1978. Ice conditions on the Chesapeake Bay as observed from Landsat during the winter of 1977. Paper read at 35th Eastern Snow Conference, at Hanover, New Hampshire.

Foster, J.L., and A.T.C. Chang. 1993. Snow Cover. In *Atlas of Satellite Observations Related to Global Change.*, edited by R.J. Gurney, J.L. Foster and C.L. Parkinson. Cambridge: Cambridge University Press.

Frei, A., and D.A. Robinson. 1999. Northern hemisphere snow extent: regional variability 1972–1994. *International Journal of Climatology* 19 (14):1535–1560.

Frezzotti, M., S. Gandolfi, F. La Marca, and S. Urbini. 2002. Snow dunes and glazed surfaces in Antarctica: new field and remote-sensing data. *Annals of Glaciology* 34:81–88.

Fricker, H.A., N.W. Young, I. Allison, and R. Coleman. 2002. Iceberg calving from the Amery Ice Shelf, East Antarctica. *Annals of Glaciology* 34:241–246.

Fujita, S., H. Maeno, T. Furukawa, and K. Matsuoka. 2002. Scattering of VHF radio waves from within the top 700 m of the Antarctic ice sheet and its relation to the depositional environment: a case-study along the Syowa-Mizuho-Dome Fuji traverse. *Annals of Glaciology* 34:157–164.

Garrity, C. 1992. Characterization of snow on floating ice and case studies of brightness temperature changes during the onset of melt. In *Microwave Remote Sensing of Sea Ice*, edited by F.D. Carsey. Washington DC: American Geophysical Union.

Garvin, J.B., and R.S. Williams. 1993. Geodetic airborne laser altimetry of Breidamerkurjökull and Skeidarárjökull, Iceland, and Jakobshavns Isbrae, West Greenland. *Annals of Glaciology* 17:379–385.

Gatto, L.W. 1990. Monitoring river ice with Landsat images. *Remote Sensing of Environment* 32 (1):1–16.

——. 1993. River ice conditions determined from ERS-1 SAR. Paper read at Eastern Snow Conference 50th Annual Meeting, at Quebec, Canada.

Gatto, L.W., and S.F. Daly. 1986. Ice conditions along the Allegheny, Monongahela and Ohio Rivers, 1983–1984. In *Internal Report*. Hanover, New Hampshire: U.S. Army Cold Regions Research and Engineering Laboratory.

Giovinetto, M.B., and H.J. Zwally. 2000. Spatial distribution of net surface accumulation on the Antarctic ice sheet. *Annals of Glaciology* 31:171–178.

Gloersen, P. 1995. Modulation of hemispheric ice cover by ENSO events. *Nature* 373:503–505.

Gloersen, P., W.J. Campbell, D.J. Cavalieri, J.C. Comiso, C.L. Parkinson, and H.J. Zwally. 1992. *Arctic and Antarctic Sea Ice, 1978–1987: Satellite Passive-Microwave Observations and Analysis*. Vol. NASA SP-511. Washington DC: National Aeronautics and Space Administration.

Gloersen, P., and D.J. Cavalieri. 1986. Reduction of weather effects in the calculation of sea ice concentrations from microwave radiances. *Journal of Geophysical Research* 91:3913–3919.

Goodison, B.E. 1989. Determination of areal snow water equivalent on the Canadian Prairies using passive microwave satellite data. Paper read at International Geoscience and Remote Sensing Symposium (IGARSS), Quantitative Remote Sensing: an Economic Tool for the Nineties, 12th Canadian Symposium on Remote Sensing, at Vancouver.

Goodison, B.E., and A.E. Walker. 1995. Canadian development and use of snow cover information from passive microwave satellite data. In *Passive Microwave Remote Sensing of Land–Atmosphere Interactions*, edited by B.J. Choudhury, Y.H. Kerr, E.G. Njoku, and P. Pampaloni. Zeist: VSP BV.

Goodison, B.E., S.E. Waterman, and E.J. Langham. 1980. Application of synthetic aperture radar data to snow cover monitoring. Paper read at Sixth Canadian Symposium on Remote Sensing, at Halifax, Nova Scotia.

Goodwin, I.D. 1990. Snow accumulation and surface topography in the katabatic zone of eastern Wilkes Land, Antarctica. *Antarctic Science* 2 (3):235–242.

Grandell, J., J.A. Johannessen, and M.T. Hallikainen. 1999. Development of a synergetic sea ice retrieval method for the ERS-1 AMI wind scatterometer and SSM/I radiometer. *IEEE Transactions on Geoscience on Remote Sensing* 37 (2):668–679.

Gratton, D.J., P.J. Howarth, and D.J. Marceau. 1990. Combining DEM parameters with Landsat MSS and TM imagery in a GIS for mountain glacier characterization. *IEEE Transactions on Geoscience and Remote Sensing* GE28 (4):766–769.

Gray, A.L., and L.D. Arsenault. 1991. Time-delayed reflections in L-band synthetic aperture radar images of icebergs. *IEEE Transactions on Geoscience and Remote Sensing* GE29:284–291.

Gray, A.L., C.E. Livingstone, and R.K. Hawkins. 1982. Testing radar systems in polar ice. *GEOS* 11:4–9.

Gray, A.L., K.E. Mattar, and P.W. Vachon. 1998. InSAR results from the RADARSAT Antarctic mapping mission data: estimation of data using a simple registration procedure. Paper read at 18th International Geoscience and Remote Sensing Symposium, at Seattle.

Gray, L., N. Short, R. Bindschadler, I. Joughin, L. Padman, P. Vornberger, and A. Khananian. 2002. RADARSAT interferometry for Antarctic grounding-zone mapping. *Annals of Glaciology* 34:269–276.

Green, R.O., J. Dozier, D. Roberts, and T. Painter. 2002. Spectral snow-reflectance models for grain-size and liquid-water fraction in melting snow for the solar-reflected spectrum. *Annals of Glaciology* 34:71–73.

Grenfell, T.C. 1983. A theoretical model of the optical properties of sea ice in the visible and near infrared. *Journal of Geophysical Research — Oceans and Atmospheres* 88 (C14):9723–9735.

Grenfell, T.C., D.J. Cavalieri, J.C. Comiso, M.R. Drinkwater, R.G. Onstott, I. Rubinstein, K. Steffen, and D.P. Winebrenner. 1992. Considerations for microwave remote sensing of thin sea ice. In *Microwave Remote Sensing of Sea Ice*, edited by F.D. Carsey. Washington DC: American Geophysical Union.

Grenfell, T.C., and G.A. Maykut. 1977. The optical properties of ice and snow in the Arctic Basin. *Journal of Glaciology* 18 (80):445–463.

Grenfell, T.C., S.G. Warren, and P.C. Mullen. 1994. Reflection of solar radiation by the Antarctic snow at ultraviolet, visible and near-infrared wavelengths. *Journal of Geophysical Research* 99 (D9):18669–18684.

Grody, N.C. 1991. Classification of snowcover and precipitation using the special sensor microwave imager. *Journal of Geophysical Research* 96:7423–7435.

Grody, N.C., and A.N. Basist. 1996. Global identification of snowcover using SSM/I measurements. *IEEE Transactions on Geoscience and Remote Sensing* 34 (1):237–249.

Gudmundsson, S., M.T. Gudmundsson, H. Björnsson, F. Sigmundsson, H. Rott, and J.M. Carstensen. 2002. Three-dimensional glacier surface motion maps at Gjálp eruption site, Iceland, inferred from combining InSAR and other displacement data. *Annals of Glaciology* 34:315–322.

Guneriussen, T. 1997. Backscattering properties of a wet snow cover derived from DEM corrected ERS-1 SAR data. *International Journal of Remote Sensing* 18 (2):375–392.

———. 1998. Snow characteristics in mountainous areas as observed with synthetic aperture radar (SAR instruments. PhD. University of Tromsø, Tromsø.

Guneriussen, T., K.A. Høgda, H. Johnson, and I. Lauknes. 2000. InSAR for estimation of changes in snow water equivalent of dry snow. Paper read at International Geosciences and Remote Sensing Symposium, at Honolulu.

Guneriussen, T., H. Johnsen, and K. Sand. 1996. DEM corrected ERS-1 SAR data for snow monitoring. *International Journal of Remote Sensing* 18 (2):181–195.

Gustajtis, K.A. 1979. Iceberg population distribution in the Labrador Sea: July data report. In *C-Core Publications*. St John's, Newfoundland: Centre for Cold Ocean Resources Engineering, Memorial University of Newfoundland.

Haefliger, M., K. Steffen, and C. Fowler. 1993. AVHRR surface temperature and narrow-band albedo comparison with ground measurements for the Greenland Ice Sheet. *Annals of Glaciology* 17:49–54.

Haefner, H., F. Holecz, E. Meier, D. Nüesch, and J. Piesbergen. 1993. Capabilities and limitations of ERS-1 SAR data for snow cover determination in mountainous regions. Paper read at Second ERS-1 Symposium: Space at the Service of Our Environment, at Hamburg, Germany.

Haefner, H., and J. Piesbergen. 1997. Methods of snow cover monitoring with active microwave data in high mountain terrain. In *Proceedings of the EARSeL Workshop Remote Sensing of Land Ice and Snow*, edited by S. Wunderle. Saint-Étienne, France: European Association of Remote Sensing Laboratories.

Hall, D.K. 1993. Active and passive microwave remote sensing of frozen lakes for regional climate studies. Glaciological data.

Hall, D.K., R.A. Bindschadler, J.L. Foster, A.T.C. Chang, and H. Siddalingaiah. 1990. Comparison of *in situ* and satellite-derived reflectances of Forbindels Glacier, Greenland. *International Journal of Remote Sensing* 11 (3):493–504.

Hall, D.K., A.T.C. Chang, J.L. Foster, C.S. Benson, and W.M. Kovalick. 1989. Comparison of in situ and Landsat derived reflectance of Alaskan glaciers. *Remote Sensing of Environment* 28:23–31.

Hall, D.K., A.T.C. Chang, and H. Siddalingaiah. 1988. Reflectances of glaciers as calculated using Landsat-5 Thematic Mapper data. *Remote Sensing of Environment* 25 (3):311–321.

Hall, D.K., D.B. Fagre, F. Klasner, G. Linebaugh, and G.E. Liston. 1994. Analysis of ERS 1 synthetic aperture radar data of frozen lakes in northern Montana and implications for climate studies. *Journal of Geophysical Research* 99 (C11):22473–22482.

Hall, D.K., J.L. Foster, A.T.C. Chang, and A. Rango. 1981. Freshwater ice thickness observations using passive microwave sensors. *IEEE Transactions on Geoscience and Remote Sensing* GE19:189–193.

Hall, D.K., J.L. Foster, J.R. Irons, and P.W. Dabney. 1993. Airborne bidirectional radiances of snow-covered surfaces in Montana, USA. *Annals of Glaciology* 17:35–40.

Hall, D.K., J.L. Foster, V.V. Salomonson, A.G. Klein, and J.Y.L. Chien. 2001. Development of a technique to assess snow-cover mapping errors from space. *IEEE Transactions on Geoscience and Remote Sensing* GE39 (2):432–438.

Hall, D.K., R.E.J. Kelly, G.A. Riggs, A.T.C. Chang, and J.L. Foster. 2002. Assessment of the relative accuracy of hemispheric-scale snow-cover maps. *Annals of Glaciology* 34:24–30.

Hall, D.K., and J. Martinec. 1985. *Remote Sensing of Ice and Snow*. London: Chapman and Hall.

Hall, D.K., J.P. Ormsby, R.A. Bindschadler, and H. Siddalingaiah. 1987. Characterization of snow and ice reflectance zones on glaciers using Landsat Thematic Mapper data. *Annals of Glaciology* 9:1–5.

Hall, D.K., G.A. Riggs, and V.V. Salomonson. 1995. Development of methods for mapping global snow cover using moderate resolution imaging spectroradiometer data. *Remote Sensing of Environment* 54:127–140.

Hall, D.K., and C. Roswell. 1981. The origin of water feeding icings on the eastern North Slope of Alaska. *Polar Record* 20:433–438.

Hall, D.K., R.S. Williams, J.S. Barton, O. Sigurdsson, L.C. Smith, and J.B. Garvin. 2000. Evaluation of remote-sensing techniques to measure decadal-scale changes of Hofsjökull ice cap, Iceland. *Journal of Glaciology* 46:375–388.

Hall, D.K., R.S. Williams, and K.J. Bayr. 1992. Glacier recession in Iceland and Austria. *Eos* 73 (12):129.

Hallikainen, M. 1986. Retrieval of the water equivalent of snow cover in Finland by satellite microwave radiometry. *IEEE Transactions on Geoscience and Remote Sensing* GE-24:855–862.

Hallikainen, M., V.I. Jääskeläinen, L. Kurvonen, J. Koskinen, E-A. Herland, and J. Perälä. 1992. Application of ERS-1 SAR data to snow mapping. Paper read at First ERS-1 Symposium: Space at the Service of Our Environment, at Cannes.

Hallikainen, M., F.T. Ulaby, and M. Abdelrazik. 1986. Dielectric properties of snow in the 3 to 37 GHz range. *IEEE Transactions on Antennas and Propagation* AP34:1329–1340.

Hallikainen, M., and D.P. Winebrenner. 1992. The physical basis for sea ice remote sensing. In *Microwave Remote Sensing of Sea Ice*, edited by F.D. Carsey. Washington, DC: American Geophysical Union.

Hallikainen, M.T. 1984. Retrieval of snow water equivalent from Nimbus-7 SMMR data: effect of land cover categories and weather conditions. *IEEE Journal of Oceanic Engineering* OE9 (5):372–376.

Hallikainen, M.T., and P.A. Jolma. 1992. Comparison of algorithms for retrieval of snow water equivalent from Nimbus-7 SMMR data in Finland. *IEEE Transactions on Geoscience and Remote Sensing* GE30 (1):124–131.

Hallikainen, M.T., F.T. Ulaby, and T.E. Van Deventer. 1987. Extinction behavior of dry snow in the 18–90 GHz range. *IEEE Transactions on Geoscience and Remote Sensing* 25:737–745.

Hambrey, M.J., and J.A. Dowdeswell. 1994. Flow regime of the Lambert Glacier–Amery Ice Shelf system, Antarctica: Structural evidence from Landsat imagery. *Annals of Glaciology* 20:401–406.

Hansen, B.U., and A. Mosbech. 1994. Use of NOAA-AVHRR data to monitor snow cover and spring meltoff in the wildlife habitats in Jameson Land, East Greenland. *Polar Research* 13 (1):125–137.

Harrison, A.R., and R.M. Lucas. 1989. Multispectral classification of snow using NOAA AVHRR imagery. *International Journal of Remote Sensing* 10 (4–5):907–916.

Haverkamp, D, L.K. Soh, and C. Tsatsoulis. 1995. A comprehensive, automated approach to determining sea ice thickness from SAR data. *IEEE Transactions on Geoscience and Remote Sensing* 33:46–57.

Hawkins, J.D., D.A. May, F. Abell, and D. Ondrejuk. 1993. Antarctic tabular iceberg A-24 movement and decay via satellite remote sensing. Paper read at Fourth International Conference on Southern Hemisphere Meteorology and Oceanography, at Hobart.

Heacock, T., T. Hirose, F. Lee, M. Manore, and B. Ramsay. 1993. Sea-ice tracking on the east coast of Canada using NOAA AVHRR imagery. *Annals of Glaciology* 17:405–413.

Herzfeld, U.C., C.S. Lingle, and L-H. Lee. 1993. Geostatistical evaluation of satellite radar altimetry for high-resolution mapping of Lambert Glacier, Antarctica. *Annals of Glaciology* 17:77–85.

Herzfeld, U.C., and M.S. Matassa. 1999. *GEOSAT Radar Altimeter DEM Atlas of Antarctica North of 72.1 Degrees South*. National Snow and Ice Data Center. (Cited 2004.) Available from http://nsidc.org/data/nsidc-0075.html.

Hewison, T.J., and S.J. English. 1999. Airborne retrievals of snow and ice surface emissivity at millimeter wavelengths. *IEEE Transactions on Geoscience and Remote Sensing* 37 (4):1871–1879.

Hiltbrunner, D., and C. Mätzler. 1997. Land surface temperature retrieval and snow discrimination using SSM/I data. Paper read at EARSeL Workshop on Remote Sensing of Land Ice and Snow, at Freiburg.

Hofer, R., and C. Mätzler. 1980. Investigations on snow parameters by radiometry in the 3- to 60-mm wavelength region. *Journal of Geophysical Research* 85 (C1):453–460.

Hoinkes, H. 1967. Glaciology in the international hydrological decade. *IAHS Commission on Snow and Ice: Reports and Discussions* 79:7–16.

Holben, B., and C.O. Justice. 1980. An examination of spectral band ratioing to reduce the topographic effect on remotely sensed data. In *Technical Memorandum*: NASA.

Holden, C. 1977. Experts ponder icebergs as relief for world water dilemma. *Science* 198:274–276.

Holladay, J.S., J.R. Rossiter, and A. Kovacs. 1990. Airborne measurement of sea ice thickness using electromagnetic induction sounding. Paper read at Ninth International Conference on Offshore Mechanical and Arctic Engineering, at Houston.

Holt, B., D.A. Rothrock, and R. Kwok. 1992. Determination of sea ice motion from satellite images. In *Microwave Remote Sensing of Sea Ice*, edited by F.D. Carsey. Washington DC: American Geophysical Union.

Hubbard, A., I. Willis, M. Sharp, D. Mair, P. Nienow, B. Hubbard, and H. Blatter. 2000. Glacier mass-balance determination by remote sensing and high-resolution modelling. *Journal of Glaciology* 46 (154):491–498.

Hufford, G.L. 1981. Sea ice detection using enhanced infrared satellite data. *Mariners Weather Log* 25 (1):1–6.

Ishizu, M., K. Mizutani, and T. Itabe. 1999. Airborne freeboard measurements of sea ice and lake ice at the Sea of Okhotsk coast in 1993–95 by a laser altimeter. *International Journal of Remote Sensing* 20 (12):2461–2476.

Jacobel, R.W., and R. Bindschadler. 1993. Radar studies at the mouths of ice streams D and E, Antarctica. *Annals of Glaciology* 17:262–268.

Jacobel, R.W., A.E. Robinson, and R.A. Bindschadler. 1994. Studies on the grounding-line location on ice streams D and E, Antarctica. *Annals of Glaciology* 20:39–42.

Jacobs, J.D., É.L. Simms, and A. Simms. 1997. Recession of the southern part of Barnes Ice Cap, Baffin Island, Canada, between 1961 and 1993, determined from digital mapping of Landsat TM. *Journal of Glaciology* 43 (143):98–102.

Jacobs, S.S., H.H. Hellmer, C.S.M. Doake, A. Jenkins, and R.M. Frohlich. 1992. Melting of ice shelves and the mass balance of Antarctica. *Journal of Glaciology* 38 (130):375–387.

Jacobsen, A., A.R. Carstensen, and J. Kamper. 1993. Mapping of satellite derived surface albedo on the Mitdluagkat Glacier, Eastern Greenland, using a digital elevation model and SPOT HRV data. *Geografisk Tidsskrift* 93:6–18.

Jeffries, M.O., K. Morris, W.F. Weeks, and H. Wakabayashi. 1994. Structural and stratigraphic features and ERS-1 SAR backscatter characteristics of ice growing on lakes in NW Alaska, winter 1991–92. *Journal of Geophysical Research* 99 (C11):22459–22471.

Jeffries, M.O., and W.M. Sackinger. 1990. Ice island detection and characterization with airborne synthetic aperture radar. *Journal of Geophysical Research* 95 (C4):5371–5377.

Jensen, H., and S. Løset. 1989. Ice management in the Barents Sea. Paper read at P.A. 89: Tenth Conference on Port and Ocean Engineering under Arctic Conditions, at Luleå.

Jezek, K. 1992. Spatial patterns in backscatter strength across the Greenland ice sheet. Paper read at First ERS-1 Symposium: Space at the Service of Our Environment, at Cannes.

Jezek, K.C. 1999. Glaciological properties of the Antarctic ice sheet from RADARSAT-1 synthetic aperture radar imagery. *Annals of Glaciology* 29:286–290.

———. 2002. RADARSAT-1 Antarctic Mapping Project: change-detection and surface velocity campaign. *Annals of Glaciology* 34:263–268.

Jin, Z., and J.J. Simpson. 1999. Bidirectional anisotropic reflectance of snow and sea ice in AVHRR channel 1 and 2 spectral regions — Part 1: theoretical analysis. *IEEE Transactions on Geoscience and Remote Sensing* 37 (1):543–554.

Jiskoot, H., P. Boyle, and T. Murray. 1998. The incidence of glacier surging in Svalbard: evidence from multivariate statistics. *Computers and Geosciences* 24 (4):387–399.

Johannessen, O.M., J.A. Johannessen, J.H. Morison, B.A. Farrelly, and E.A.S. Svendsen. 1983. Oceanographic conditions in the marginal ice zone north of Svalbard in early fall 1979 with emphasis on mesoscale processes. *Journal of Geophysical Research* 88:2755–2769.

Johannessen, O.M., S. Sandven, and K. Kloster. 1991. Remote sensing of icebergs in the Barents Sea during SI.E. 89. Paper read at POAC 91: Eleventh International Conference on Port and Ocean Engineering under Arctic Conditions, at St John's, Newfoundland.

Johannessen, O.M., E.V. Shalina, and M.W. Miles. 1999. Satellite evidence for an Arctic sea ice cover in transformation. *Science* 286 (5446):1937–1939.

Josberger, E.G., and N.M. Mognard. 1998. A passive microwave snow-depth algorithm with a proxy for snow metamorphism. Paper read at Fourth International Workshop on Applications of Remote Sensing in Hydrology, at Santa Fe.

Josberger, E.G., N.M. Mognard, B. Lind, R. Matthews, and T. Carroll. 1998. Snowpack water-equivalent estimates from satellite and aircraft remote-sensing measurements of the Red River basin, north-central U.S.A. *Annals of Glaciology* 26:119–124.

Joughin, I. 2002. Ice-sheet velocity mapping: a combined interferometric and speckle-tracking approach. *Annals of Glaciology* 34:195–201.

Joughin, I., D. Winebrenner, M. Fahnestock, R. Kwok, and W. Krabill. 1996. Measurement of ice-sheet topography using satellite-radar interferometry. *Journal of Glaciology* 42:10–22.

Joughin, I.R., R. Kwok, and M.A. Fahnestock. 1996. Estimation of ice sheet motion using satellite radar interferometry: Method and error analysis with application to Humboldt Glacier, Greenland. *Journal of Glaciology* 42 (142):564–575.

———. 1998. Interferometric estimation of three-dimensional ice-flow using ascending and descending passes. *IEEE Transactions on Geoscience and Remote Sensing* 36 (1):25–37.

Kääb, A., F. Paul, M. Maisch, M. Hoelzle, and W. Haeberli. 2002. The new remote-sensing-derived Swiss glacier inventory: II: First results. *Annals of Glaciology* 34:362–366.

Kapitsa, A.P., J.K. Ridley, G.deQ. Robin, M.J. Siegert, and I.A. Zotikov. 1996. A large deep freshwater lake beneath the ice of central East Antarctica. *Nature* 381:684–686.

Kargel, J.S. 2000. New eyes in the sky measure glaciers and ice sheets. *Eos* 81 (24):265, 270–271.

Kelly, R.E.J. 2002. Estimation of the ELA on Hardangerjøkulen, Norway, during the 1995/96 winter season using repeat-pass SAR coherence. *Annals of Glaciology* 34:349–354.

Kendra, J.R., K. Sarabandi, and F.T. Ulaby. 1998. Radar measurements of snow: experiment and analysis. *IEEE Transactions on Geoscience and Remote Sensing* 36 (3):864–879.

Kennett, M., and T. Eiken. 1997. Airborne measurements of glacier surface elevation by scanning laser altimeter. *Annals of Glaciology* 24:293–296.

Ketchum, R.D. 1971. Airborne laser profiling of the Arctic packice. *Remote Sensing of Environment* 2:41–52.

Key, J., and M. Haefliger. 1992. Arctic ice surface temperature retrieval from AVHRR thermal channels. *Journal of Geophysical Research* 97 (D5):5885–5893.

Keys, H.J.R., S.S. Jacobs, and L.W. Brigham. 1998. Continued northward expansion of the Ross Ice Shelf, Antarctica. *Annals of Glaciology* 27:93–98.

Keys, J.R., S.S. Jacobs, and D. Barnett. 1990. The calving and drift of iceberg B-9 in the Ross Sea, Antarctica. *Antarctic Science* 2:243–257.

Khvorostovsky, K.S., L.P. Bobylev, and O.M. Johannessen. 2003. Greenland ice sheet elevation variations. In *Arctic Environment Variability in the Context of Global Change*, edited by L.P. Bobylev, K.Y. Kondratyev, and O.M. Johannessen. Chichester: Praxis-Springer.

Kidder, S.Q., and H.T. Wu. 1984. Dramatic contrast between low cloud and snow cover in daytime 3.7 μm images. *Monthly Weather Review* 112 (11):2345–2346.

Kirby, M.E. 1982. Digital image analysis of SAR imagery for the detection of icebergs. *Iceberg Research* 2:6–18.

Kirby, M.E., and R.J. Lowry. 1979. Iceberg detectability problems using SAR and SLAR systems. Paper read at Fifth Annual W.T. Pecora Symposium: Satellite Hydrology, at Sioux Falls, South Dakota.

Klein, A.G., and D.K. Hall. 1999. Snow albedo determination using the NASA MODIS instrument. Paper read at Eastern Snow Conference, 55th Annual Meeting, at Fredericton, New Brunswick.

Klein, A.G., D.K. Hall, and G.A. Riggs. 1998. Improving snow-cover mapping in forests through the use of a canopy reflectance model. *Hydrological Processes* 12:1723–1744.

Klein, A.G., and J. Stroeve. 2002. Development and validation of a snow albedo algorithm for the MODIS instrument. *Annals of Glaciology* 34:45–52.

Kloster, K., and W. Spring. 1993. Iceberg and glacier mapping using satellite optical imagery during the Barents Sea ice surface data acquisition program (IDAP). Paper read at POAC 93: Twelfth International Conference on Port and Ocean Engineering under Arctic Conditions, at Hamburg.

Knap, W.H., and J. Oerlemans. 1996. The surface albedo of the Greenland ice sheet: satellite-derived and in situ measurements in the Sondre Stromfjord area during the 1991 melt season. *Journal of Glaciology* 42 (141):364–374.

Knap, W.H., and C.H. Reijmer. 1998. Anisotropy of the reflected radiation field over melting glacier ice: Measurements in Landsat TM bands 2 and 4. *Remote Sensing of Environment* 65:93–104.

Knap, W.H., C.H. Reijmer, and J. Oerlemans. 1999. Narrowband to broadband conversion of Landsat TM glacier albedos. *International Journal of Remote Sensing* 20 (10):2091–2110.

Koelemeijer, R., J. Oerlemans, and S. Tjemkes. 1993. Surface reflectance of Hinteresiferner, Austria, from Landsat 5 TM imagery. *Annals of Glaciology* 17:17–22.

Koenig, L.S., K.R. Greenaway, M. Dunbar, and G. Hattersley-Smith. 1952. Arctic ice islands. *Arctic* 5:67–103.

Koerner, R.M. 1970. Some observations on superimposition of ice on the Devon Island ice cap. *Geographical Annals* 52a (1):57–67.

———. 1989. Ice core evidence for extensive melting of the Greenland ice sheet in the last interglacial. *Science* 244 (4907):964–968.

Komyshenets, V.I., and Ye.B. Leont'yev. 1989. The drift of a giant iceberg in the Weddell Sea. [Dreyf gigantskogo aysberga v more Uedella.] *Polar Geography and Geology* 13 (1):68–71.

König, M., J. Wadham, J-G. Winther, J. Kohler, and A-M. Nuttall. 2002. Detection of superimposed ice on the glaciers Kongsvegen and midre Lovénbreen, Svalbard, using SAR imagery. *Annals of Glaciology* 34:335–342.

König, M., J-G. Winther, and E. Isaksson. 2001. Measuring snow and glacier ice properties from satellite. *Reviews of Geophysics* 39 (1):1–27.

Korsnes, R., S.R. Souza, R. Donangelo, A. Hansen, M. Paczuski, and K. Sneppen. 2004. Scaling in fracture and refreezing of sea ice. *Physica A* 331:291–296.

Koskinen, J., L. Kurvonen, V. Jääskeläinen, and M. Hallikainen. 1994. Capability of radar and microwave radiometer to classify snow types in forested areas. Paper read at International Geoscience and Remote Sensing Symposium. Surface and Atmospheric Remote Sensing: Technologies, Data Analysis and Interpretation, at Pasadena.

Koskinen, J.T., J.T. Pulliainen, and M. Hallikainen. 1997. The use of ERS-1 SAR data in snow melt monitoring. *IEEE Transactions on Geoscience and Remote Sensing* 35 (3):601–610.

Kotlyakov, V.M. 1970. Land glaciation part in the earth's water balance. Paper read at IAHS/Unesco Symposium on World Water Balance, at Reading.

Kovacs, A., A.J. Gow, and R.M. Morey. 1995. The in-situ dielectric constant of polar firn revisited. *Cold Regions Science and Technology* 23 (3):245–256.

Kovacs, A., and R.M. Morey. 1986. Electromagnetic measurements of multiyear sea ice using impulse radar. *Cold Regions Science and Technology* 12:67–93.

Krabill, W., and 9 others. 2000. Greenland ice sheet: high-elevation balance and peripheral thinning. *Science* 289 (5478):428–430.

Krabill, W.B., R.N. Swift, and W.B. Tucker. 1990. Recent measurements of sea ice topography in the Eastern Arctic. In *Sea Ice Properties and Processes*, edited by S.F. Ackley and W.F. Weeks. Hanover, New Hampshire: U.S. Army Cold Regions Research and Engineering Laboratory.

Kramer, H.J. 1996. *Observation of the Earth and Its Environment*. 3rd ed. Berlin: Springer.

Krimmel, R.M., and M.F. Meier. 1975. Glacier applications of ERTS images. *Journal of Glaciology* 15 (73):391–402.

Kuga, Y., F.T. Ulaby, T.F. Haddock, and R.D. Deroo. 1991. Millimeter-wave radar scattering from snow.1. Radiative-transfer model. *Radio Science* 26 (2):329–341.

Kuittinen, R. 1997. Optical and thermal sensors in snow cover modelling. Paper read at EARSeL Workshop on Remote Sensing of Land Ice and Snow, at Freiburg.

Kurvonen, L., and M. Hallikainen. 1997. Influence of land-cover category on brightness temperature of snow. *IEEE Transactions on Geoscience and Remote Sensing* 35 (2):367–377.

Kwok, R. 2002. Arctic sea-ice area and volume production: 1996/97 versus 1997/98. *Annals of Glaciology* 34:447–453.

Kwok, R., G. Cunningham, and B. Holt. 1992. An approach to the identification of sea ice types from spaceborne SAR data. In *Microwave Remote Sensing of Sea Ice*, edited by F.D. Carsey. Washington DC: American Geophysical Union.

Kwok, R., and G.F. Cunningham. 1994. Backscatter characteristics of the winter ice cover in the Beaufort Sea. *Journal of Geophysical Research* 99:7787–7802.

Kwok, R., J.C. Curlander, R. McConnell, and S.S. Pang. 1990. An ice-motion tracking system at the Alaska SAR Facility. *IEEE Journal of Oceanic Engineering* OE15 (1):44–54.

Kwok, R., and M.A. Fahnestock. 1996. Ice sheet motion and topography from radar interferometry. *IEEE Transactions on Geoscience and Remote Sensing* 34:189–200.

Kwok, R., E. Rignot, B. Holt, and R.G. Onstott. 1992. Identification of sea ice type in spaceborne SAR data. *Journal of Geophysical Research* 97:2391–2402.

Laberge, M.J., and S. Payette. 1995. Long-term monitoring of permafrost change in palsa peatland in northern Quebec, Canada: 1983–1993. *Arctic and Alpine Research* 27:167–171.

Lachenbruch, A.H., and B.V. Marshall. 1986. Changing climate: geothermal evidence from permafrost in the Alaskan Arctic. *Science* 234:689–696.

Larrowe, B.T., R.B. Innes, R.A. Rendleman, and R.J. Porcello. 1971. Lake ice surveillance via airborne radar: some experimental results. Ann Arbor, Michigan: University of Michigan.

Larson, R.W., R.A. Schuchman, R.A. Rawson, and R.D. Worsfold. 1978. The use of SAR systems for iceberg detection and characterization. Paper read at Twelfth International Symposium on Remote Sensing of Environment, at Ann Arbor.

Laxon, S.W.C. 1989. Satellite radar altimetry over sea ice. PhD. Mullard Space Science Laboratory, University College London, London.

Lazzara, M.A., K.C. Jezek, T.A. Scambos, D.R. MacAyeal, and C.J. Van der Veen. 1999. On the recent calving of icebergs from the Ross Ice Shelf. *Polar Geography* 23 (3):201–212.

Leconte, R., and P.D. Klassen. 1991. Lake and river ice investigations in northern Manitoba using airborne SAR imagery. *Arctic* 44 (supp 1):153–163.

Lefauconnier, B., J.O. Hagen, and J.P. Rudant. 1994. Flow speed and calving rate of Kongsbreen glacier, Svalbard, using SPOT images. *Polar Research* 13 (1):59–65.

Leshkevich, G.A. 1981. Categorization of Northern Green Bay ice cover using Landsat-1 digital data — a case study. In *NOAA Technical Memorandum*: National Oceanographic and Atmospheric Administration.

———. 1985. Machine classification of freshwater ice types from Landsat-1 digital data using ice albedos as training sets. *Remote Sensing of Environment* 17:251–263.

Lewis, E.O., C.E. Livingstone, C. Garrity, and J.R. Rossiter. 1994. Properties of ice and snow. In *Remote Sensing of Sea Ice and Icebergs*, edited by S. Haykin, E.O. Lewis, R.K. Raney, and J.R. Rossiter. New York: John Wiley & Sons.

Li, Z.Q., and H.G. Leighton. 1992. Narrow-band to broad-band conversion with autocorrelated reflectance measurements. *Journal of Applied Meteorology* 31 (5):421–432.

Liu, A.K., and D.J. Cavalieri. 1998. On sea ice drift from the wavelet analysis of the Defense Meteorological Satellite Program (DMSP Special Sensor Microwave Imager (SSM/I) data. *International Journal of Remote Sensing* 19 (7):1415–1423.

Liu, H., K.C. Jezek, and B. Li. 1999. Development of an Antarctic digital elevation model by integrating cartographic and remotely sensed data: a geographic information system based approach. *Journal of Geophysical Research* 104 (B10):23199–23213.

Livingstone, C.E., R.K. Hawkins, A.L. Gray, L.D. Arsenault, K. Okamoto, T.L. Wilkinson, and D. Pearson. 1983. The CCRS/SURSAT active–passive experiment 1978–1980. The microwave signatures of sea ice. Ottawa: Canada Centre for Remote Sensing.

Livingstone, C.E., R.G. Onstott, L.D. Arsenault, A.L. Gray, and K.P. Singh. 1987. Microwave sea-ice signatures near the onset of melt. *IEEE Transactions on Geoscience and Remote Sensing* 25:174–187.

Løset, S., and T. Carstens. 1993. Production of icebergs and observed extreme drift speeds in the Barents Sea. Paper read at POAC 93: Twelfth International Conference on Port and Ocean Engineering under Arctic Conditions, at Hamburg.

Løvås, S.M., W. Spring, and A. Holm. 1993. Stereo photogrammetric analysis of icebergs and sea ice from the Barents Sea ice data acquisition program (IDAP). Paper read at POAC 93: Twelfth International Conference on Port and Ocean Engineering under Arctic Conditions, at Hamburg.

Lowry, R.T., and J. Miller. 1983. Iceberg mapping in Lancaster Sound with synthetic aperture radar. *Iceberg Research* 6:3–9.

Lucchitta, B.K., K.F. Mullins, A.L. Allison, and J.G. Ferrigno. 1993. Antarctic glacier-tongue velocities from Landsat images: first results. *Annals of Glaciology* 17:356–366.

Lucchitta, B.K., C.F. Rosanova, and K.F. Mullins. 1995. Velocities of Pine Island Glacier, West Antarctica, from ERS-1 SAR images. *Annals of Glaciology* 21:277–283.

Lundstrom, S.C., A.E. McCafferty, and J.A. Coe. 1993. Photogrammetric analysis of 1984–89 surface altitude change of the partially debris-covered Eliot Glacier, Mount Hood, Oregon, USA. *Annals of Glaciology* 17:167–170.

Lure, Y.M.F., N.C. Grody, H.Y.M. Yeh, and J.S.J. Lin. 1992. Neural network approaches to classification of snow cover and precipitation from special sensor microwave imager (SSM/I). Paper read at Eighth International Conference on Interactive Information and Processing Systems (IIPS) for Meteorology, Oceanography and Hydrology, at Atlanta, Georgia.

Lythe, M., A. Hauser, and G. Wendler. 1999. Classification of sea ice types in the Ross Sea, Antarctica from SAR and AVHRR imagery. *International Journal of Remote Sensing* 20 (15):3073–3085.

Mackay, D.K., and O.H. Løken. 1974. Arctic hydrology. In *Arctic and Alpine Environments*, edited by J.D. Ives and R.G. Barry. London: Methuen.

Macqueen, A.D. 1988. Radio echo-sounding as a glaciological technique: a bibliography. Cambridge: World Data Centre "C" for glaciology.

Markus, T., D.J. Cavalieri, and A. Ivanoff. 2002. The potential of using Landsat 7 ETM+ for the classification of sea-ice surface conditions during summer. *Annals of Glaciology* 34:415–419.

Marshall, G.J., W.G. Rees, and J.A. Dowdeswell. 1993. Limitations imposed by cloud cover on multitemporal visible sand satellite data sets from polar regions. *Annals of Glaciology* 17:113–120.

Marshall, G.J., W.G. Rees, and J.A. Dowdeswell. 1995. The discrimination of glacier facies in ERS-1 SAR data. In *Sensors and Environmental Applications of Remote Sensing Data*, edited by J. Askne. Rotterdam: A.A. Balkema.

Martin, S., K. Steffen, J. Comiso, D. Cavalieri, M.R. Drinkwater, and B. Holt. 1992. Microwave remote sensing of polynyas. In *Microwave Remote Sensing of Sea Ice*, edited by F.D. Carsey. Washington DC: American Geophysical Union.

Martin, T.V., H.J. Zwally, A.C. Brenner, and R.A. Bindschadler. 1983. Analysis and retracking of continental ice sheet radar altimeter waveforms. *Journal of Geophysical Research* 88:1608–1616.

Martinec, J. 1977. Expected snow loads on structures from incomplete hydrological data. *Journal of Glaciology* 19 (81):185–195.

Martinec, J., A. Rango, and E. Major. 1983. The Snowmelt Runoff-Model (SRM) user's manual. In *NASA Reference Publication*: National Aeronatics and Space Administration.

Maslanik, J.A., and R.G. Barry. 1987. Lake ice formation and breakup as an indicator of climatic change potential for monitoring remote sensing techniques. Paper read at The Influence of Climatic Change and Climatic Variability on the Hydrologic Regime and Water Resources, at Vancouver.

Massom, R.A. 1991. *Satellite Remote Sensing of Polar Regions*. Boca Raton, Florida: Lewis Publications.

Matson, M., C.F. Ropelewski, and M.S. Varnardore. 1986. *An Atlas of Satellite-Derived Northern Hemisphere Snow Cover Frequency*. Washington, DC: U.S. Department of Commerce. National Oceanic and Atmospheric Administration Data and Information Service. National Environmental Satellite, Data and Information Service.

Matsuoka, K., H. Maeno, S. Uratsuka, S. Fujita, T. Furukawa, and O. Watanabe. 2002. A ground-based, multi-frequency ice-penetrating radar system. *Annals of Glaciology* 34:171–176.

Matsuoka, T., S. Uratsuka, M. Satake, A. Nadai, T. Umehara, H. Maeno, H. Wakabayashi, F. Nishio, and Y. Fukamachi. 2002. Deriving sea-ice thickness and ice types in the Sea of Okhotsk using dual-frequency airborne SAR (Pi-SAR) data. *Annals of Glaciology* 34:429–434.

Mätzler, C. 1987. Applications of the interaction of microwaves with the natural snow cover. *Remote Sensing Reviews* 2:259–387.

——. 1994. Passive microwave signatures of landscapes in winter. *Meteorology and Atmospheric Physics* 54 (1–4):241–260.

Mätzler, C., and U. Wegmüller. 1987. Dielectric properties of freshwater ice at microwave frequencies. *Journal of Physics D: Applied Physics* 20:1623–1630.

Maxfield, A.W. 1994. Radar satellite snowmelt detection in the Canadian Rocky Mountains. Paper read at International Geoscience and Remote Sensing Symposium. Surface and Atmospheric Remote Sensing: Technologies, Data Analysis and Interpretation, at Pasadena.

Maykut, G.A. 1986. The surface heat and mass balance. In *Geophysics of Sea Ice*, edited by N. Untersteiner. London: Plenum Press.

Maykut, G.A. 1985. The ice environment. In *Sea Ice Biota*, edited by R.A. Horner. Boca Raton, Florida: CRC Press.

Meier, M.F. 1984. Contribution of small glaciers to global sea level. *Science* 226 (4681):1418–1421.

———. 1990. Reduced rise in sea level. *Nature* 343:115.

Melling, H. 1998. Detection of features in first-year pack ice by synthetic aperture radar (SAR). *International Journal of Remote Sensing* 19 (6):1223–1249.

Melloh, R.A., and L.W. Gatto. 1990. Interpretation of passive and active microwave imagery over snow-covered lakes and rivers near Fairbanks, Alaska. Paper read at Workshop on Applications of Remote Sensing in Hydrology, at Saskatoon, Saskatchewan.

Merson, R.H. 1989. An AVHRR mosaic image of Antarctica. *International Journal of Remote Sensing* 10:669–674.

Michel, B. 1971. Winter regime of rivers and lakes. In *Cold Regions Science and Engineering Monographs*. Hanover, New Hampshire: U.S. Army Cold Regions Research and Engineering Laboratory.

Michel, R., and E. Rignot. 1999. Flow of Glacier Moreno, Argentina, from repeat-pass Shuttle Imaging Radar images: comparison of the phase correlation method with radar interferometry. *Journal of Glaciology* 45 (149):93–100.

Middleton, W.E., and A.G. Mungall. 1952. The luminous directional reflectance of snow. *Journal of the Optical Society of America* 42:572–579.

Mognard, N. 2003. Snow cover dynamics. In *Arctic Environment Variability in the Context of Global Change*, edited by L.P. Bobylev, K.Y. Kondratyev, and O.M. Johannessen. Chichester: Praxis-Springer.

Mognard, N.M., and E.G. Josberger. 2002. Northern Great Plains 1996/97 seasonal evolution of snowpack parameters from satellite passive-microwave measurements. *Annals of Glaciology* 34:15–23.

Morris, K., M.O. Jeffries, and W.S. Weeks. 1995. Ice processes and growth history on Arctic and sub-Arctic lakes using ERS-1 SAR data. *Polar Record* 31 (117):115–128.

Mote, T.L., and M.R. Anderson. 1995. Variations in snowpack melt on the Greenland ice sheet based on passive-microwave measurements. *Journal of Glaciology* 41 (137):51–60.

Mote, T.L., M.R. Anderson, K.C. Kuivinen, and C.M. Rowe. 1993. Passive microwave-derived spatial and temporal variations of summer melt on the Greenland ice sheet. *Annals of Glaciology* 17:233–238.

Müller, F. 1962. Zonation in the accumulation area of the glaciers of Axel Heiberg Island, NWT. Canada. *Journal of Glaciology* 4:302–313.

Müller, F., T. Caflisch, and G. Müller. 1976. *Firn und Eis der Schweitzer Alpen: Gletscherinventar, Geographisches Institut Publ.* Zürich: Eidgenössische Technische Hochschule.

Murphy, D.L., and G.L. Wright. 1991. Iceberg movement determined by satellite tracked platforms: International Ice Patrol in the North Atlantic.

Murray, T., T. Strozzi, A. Luckman, H. Pritchard, and H. Jiskoot. 2002. Ice dynamics during a surge of Sortebrae, East Greenland. *Annals of Glaciology* 34:323–329.

Nagler, T. 1991. Verfahren zur Analyse der Schneebedeckung aus Messungen des SSM/I. Diplomarbeit, Universität Innsbruck, Innsbruck.

Nagler, T., and H. Rott. 1997. The application of ERS-1 SAR for snowmelt runoff modelling. Paper read at Remote Sensing and Geographic Information Systems for Design and Operation of Water Resources Systems, at Rabat, Morocco.

Nagurny, A.P., V.G. Korostelev, and V.V. Ivanov. 1999. Multiyear variability of sea ice thickness in the Arctic Basin measured by elastic-gravity waves on the ice surface. *Meteorologiya i Hidrologiya* 3:72–78.

Narayanan, R.M., and S.R. Jackson. 1994. Snow cover classification using millimeter-wave radar imagery. Paper read at International Geoscience and Remote Sensing Symposium. Surface and Atmospheric Remote Sensing: Technologies, Data Analysis and Interpretation, at Pasadena.

Negri, A.J., E.J. Nelkin, R.F. Adler, G.J. Huffman, and C. Kummerow. 1995. Evaluation of passive microwave precipitation algorithms in wintertime midlatitude situations. *Journal of Atmospheric and Oceanic Technology* 12:20–32.

Ninnis, R.M., W.J. Emery, and M.J. Collins. 1986. Automated extraction of pack ice motion from advanced very high resolution radiometer imagery. *Journal of Geophysical Research* 91 (C9):10725–10734.

Nolin, A.W., and J. Dozier. 1993. Estimating snow grain-size using AVIRIS data. *Remote Sensing of Environment* 44 (2–3):231–238.

———. 2000. A hyperspectral method for remotely sensing the grain size of snow. *Remote Sensing of Environment* 74 (2):207–216.

Nolin, A.W., J. Dozier, and L.A.K. Mertes. 1993. Mapping alpine snow cover using a spectral mixture modeling technique. *Annals of Glaciology* 17:121–124.

Nolin, A.W., and J. Stroeve. 1997. The changing albedo of the Greenland ice sheet: implications for climate modeling. *Annals of Glaciology* 25:51–57.

Nyfors, E. 1982. On the dielectric properties of dry snow in the 800 MHz to 13 GHz range. Helsinki: Radio Laboratory, Helsinki University of Technology.

Nystuen, J.A., and F.W. Garcia. 1992. Sea ice classification using SAR backscatter statistics. *IEEE Transactions on Geoscience and Remote Sensing* GE30:502–509.

Onstott, R.G. 1992. SAR and scatterometer signatures of sea ice. In *Microwave Remote Sensing of Sea Ice*, edited by F.D. Carsey. Washington DC: American Geophysical Union.

Oppenheimer, M. 1998. Global warming and the stability of the West Antarctic ice sheet. *Nature* 393 (6683):325–332.

Orheim, O. 1988. Antarctic icebergs — production, distribution and disintegration. *Annals of Glaciology* 11:205.

Orheim, O., and B.K. Lucchitta. 1988. Numerical analysis of Landsat thematic mapper images of Antarctica: Surface temperatures and physical properties. *Annals of Glaciology* 11:109–120.

Padman, L., H.A. Fricker, R. Coleman, S. Howard, and L. Erofeeva. 2002. A new tide model for the Antarctic ice shelves and seas. *Annals of Glaciology* 34:247–254.

Palecki, M.A., and R.G. Barry. 1986. Freeze-up and break-up of lakes as an index of temeparture changes during the transition seasons: a case study in Finland. *Journal of Climate and Applied Meteorology* 25:893–902.

Papa, F., B. Legresy, N.M. Mognard, E.G. Josberger, and F. Rémy. 2002. Estimating terrestrial snow depth with Topex-Poseidon altimeter and radiometer. *IEEE Transactions on Geoscience and Remote Sensing* 40 (10):2162–2169.

Parashar, S.K., C. Roche, and R.D. Worsfold. 1978. Four channel synthetic aperture radar imagery results of freshwater ice and sea ice in Lake Melville. St John's, Newfoundland: C-Core.

Paren, J., and G. deQ. Robin. 1975. Internal reflections in polar ice sheets. *Journal of Glaciology* 14 (71):251–259.

Parkinson, C.L. 1994. Spatial patterns in the length of the sea ice season in the Southern Ocean, 1979–1986. *Journal of Geophysical Research* 99 (C8):16327–16339.

———. 2002. Trends in the length of the Southern Ocean sea-ice season, 1979–99. *Annals of Glaciology* 34:435–440.

Parkinson, C.L., and D.J. Cavalieri. 2002. A 21 year record of Arctic sea-ice extents and their regional, seasonal and monthly variability trends. *Annals of Glaciology* 34:441–446.

Parkinson, C.L., D.J. Cavalieri, P. Gloersen, H.J. Zwally, and J. Comiso. 1999. Arctic sea ice extents, areas and trends, 1978–1996. *Journal of Geophysical Research* 104 (C9):20837–20856.

Parkinson, C.L., and P. Gloersen. 1993. Global sea ice coverage. In *Atlas of Satellite Observations Related to Global Change*, edited by R.J. Gurney, J.L. Foster, and C.L. Parkinson. Cambridge: Cambridge University Press.

Parrot, J.F., N. Lyberis, B. Lefauconnier, and G. Manby. 1993. SPOT multispectral data and digital terrain model for the analysis of ice-snow fields on Antarctic glaciers. *International Journal of Remote Sensing* 14 (3):425–440.

Partington, K.C. 1998. Discrimination of glacier facies using multi-temporal SAR data. *Journal of Glaciology* 44 (146):42–53.

Paterson, W.S.B. 1994. *The Physics of Glaciers*, 3rd ed. Kidlington: Elsevier Science.

Pattyn, F., and H. Decleir. 1993. Satellite monitoring of ice and snow conditions in the Sør Rondane Mountains, Antarctica. *Annals of Glaciology* 17:41–48.

Paul, F., A. Kääb, M. Maisch, T. Kellenberger, and W. Haeberli. 2002. The new remote-sensing-derived Swiss glacier inventory: I: Methods. *Annals of Glaciology* 34:355–361.

Peel, D.A. 1992. Spatial temperature and accumulation rate variations in the Antarctic Peninsula. In *The Contribution of Antarctic Peninsula Ice to Sea Level Rise (CEC Project Report EPOC-CT90-0015)*, edited by E.M. Morris. Cambridge: British Antarctic Survey.

Perovich, D.K. 1989. A two-stream multilayer, spectral radiative transfer model for sea ice. Hanover, New Hampshire: Cold Regions Research and Engineering Laboratory.

Perovich, D.K., and T.C. Grenfell. 1981. Laboratory studies of the optical properties of young sea ice. *Journal of Glaciology* 96:331–346.

Perovich, D.K., G.A. Maykut, and T.C. Grenfell. 1986. Optical properties of ice and snow in the polar oceans. I: Observations. *SPIE Journal* 637:232–244.

Pivot, F.C., C. Kergomard, and C.R. Duguay. 2002. Use of passive-microwave data to monitor spatial and temporal variations of snow cover at tree line near Churchill, Manitoba, Canada. *Annals of Glaciology* 34:58–64.

Proy, C., D. Tanré, and P.Y. Deschamps. 1989. Evaluation of topographic effects in remotely sensed data. *Remote Sensing of Environment* 30:21–32.

Pulliainen, J.T., and M. Hallikainen. 2001. Retrieval of Regional Snow Water Equivalent from Space-Borne Passive Microwave Observations. *Remote Sensing of Environment* 75 (1):76–85.

Qunzhu, Z., C. Meishing, and F. Xuezhi. 1984. Study on Spectral reflection characteristics of snow, ice and water of northwestern China. *Sci. Sinisu (Series B)* 27:647–656.

Ramage, J.M., and B.L. Isacks. 2002. Determination of melt-onset and refreeze timing on southeast Alaskan icefields using SSM/I diurnal amplitude variations. *Annals of Glaciology* 34:391–398.

Ramage, J.M., B.L. Isacks, and M.M. Miller. 2000. Radar glacier zones in southeast Alaska, USA: field and satellite observations. *Journal of Glaciology* 46 (153):287–296.

Ramsay, B.H. 1998. The interactive multisensor snow and ice mapping system. *Hydrological Processes* 12 (10–11):1537–1546.

Rango, A. 1993. Snow hydrology processes and remote-sensing. 2. *Hydrological Processes* 7 (2):121–138.

Rango, A., A.T.C. Chang, and J.L. Foster. 1979. The utilization of spaceborne microwave radiometers for monitoring snowpack properties. *Nordic Hydrology* 10 (1):25–40.

Rango, A., and A.I. Shalaby. 1998. Operational applications of remote sensing in hydrology: success, prospects and problems. *Hydrological Sciences Journal (Journal Des Sciences Hydrologiques)* 43 (6):947–968.

Rau, F., and M. Braun. 2002. The regional distribution of the dry-snow zone on the Antarctic Peninsula north of 70° S. *Annals of Glaciology* 34:95–100.

Rees, W.G. 1988. Synthetic aperture radar data over terrestrial ice. Paper read at International Geoscience and Remote Sensing Symposium. Remote Sensing: Moving Towards the 21st Century, at Edinburgh.

——. 1999. *The Remote Sensing Data Book*. Cambridge: Cambridge University Press.

——. 2001. *Physical Principles of Remote Sensing*, 2nd ed. Cambridge: Cambridge University Press.

Rees, W.G., and R.E. Donovan. 1992. Refraction correction for radio echo sounding of large ice masses. *Journal of Glaciology* 38 (129):302–308.

Rees, W.G., and J.A. Dowdeswell. 1988. Topographic effects on light scattering from snow. Paper read at IGARSS 88, at Edinburgh.

Rees, W.G., J.A. Dowdeswell, and A.D. Diament. 1995. Analysis of ERS-1 Synthetic Aperture Radar data from Nordaustlandet, Svalbard. *International Journal of Remote Sensing* 16 (5):905–924.

Rees, W.G., and I. Lin. 1993. Texture-based classification of cloud and ice-cap surface features. *Annals of Glaciology* 17:250–254.

Rees, W.G., and A.M. Steel. 2001. Radar backscatter coefficients and snow detectability for upland terrain in Scotland. *International Journal of Remote Sensing* 22 (15):3015–3026.

Richards, J.A. 1993. *Remote Sensing Digital Image Analysis*. 2nd ed. Berlin: Springer-Verlag.

Ridley, J.K. 1993a. Climate signals from SSM/I observations of marginal ice shelves. *Annals of Glaciology* 17:189–194.

——. 1993b. Surface melting on Antarctic Peninsula ice shelves detected by passive microwave sensors. *Geophysical Research Letters* 20 (23):2639–2642.

Ridley, J.K., and K.C. Partington. 1988. A model of satellite radar altimeter return from ice sheets. *International Journal of Remote Sensing* 9:601–624.

Riggs, G.A., D.K. Hall, and S.A. Ackerman. 1999. Sea ice extent and classification mapping with the moderate resolution imaging spectroradiometer airborne simulator. *Remote Sensing of Environment* 68 (2):152–163.

Rignot, E. 2002. Mass balance of East Antarctic glaciers and ice shelves from satellite data. *Annals of Glaciology* 34:217–227.

Rignot, E., G. Buscarlet, B. Csatho, S. Gogineni, W. Krabill, and M. Schmeltz. 2000. Mass balance of the northeast sector of the Greenland ice sheet: a remote-sensing perspective. *Journal of Glaciology* 46 (153):265–273.

Rivera, A., C. Acuña, G. Casassa, and F. Bown. 2002. Use of remotely sensed and field data to estimate the contribution of Chilean glaciers to eustatic sea-level rise. *Annals of Glaciology* 34:367–372.

Robin, G. deQ., S. Evans, and J.T. Bailey. 1969. Interpretation of radio echo sounding in polar ice sheets. *Philosophical Transactions of the Royal Society of London Series A — Mathematical Physical and Engineering Sciences* 265 (1166):437–505.

Robinson, D.A. 1993. Hemispheric snow cover from satellites. *Annals of Glaciology* 17:367–371.

———. 1997. Hemispheric snow cover and surface albedo for model validation. *Annals of Glaciology* 25:241–245.

———. 1999. Northern hemisphere snow cover during the satellite era. Paper read at Fifth Conference on Polar Meteorology and Oceanography, at Dallas, Texas.

Robinson, D.A., K.F. Dewey, and R.R. Heim. 1993. Global snow cover monitoring: an update. *Bulletin of the American Meteorological Society* 74 (9):1689–1696.

Romanov, P., G. Gutman, and I. Csiszar. 2000. Automated monitoring of snow cover over North America using multispectral satellite data. *Journal of Applied Meteorology* 39:1866–1890.

Rosenfeld, S., and N. Grody. 2000. Metamorphic signature of snow revealed in SSM/I measurements. *IEEE Transactions on Geoscience and Remote Sensing* 38 (1):53–63.

Rosenthal, W., and J. Dozier. 1996. Automated mapping of montane snow cover at sub-pixel resolution from Landsat Thematic Mapper. *Water Resources Research* 6 (12):2370–2393.

Ross, B., and J. Walsh. 1986. Synoptic-scale influences of snow cover and sea ice. *Monthly Weather Review* 114 (10):1795–1810.

Rossiter, J.R., L.D. Arsenault, J. Benoit, A.L. Gray, E.V. Guy, D.J. Lapp, R.O. Ramseier, and E. Wedler. 1984. Detection of icebergs by airborne imaging radars. Paper read at Ninth Canadian Symposium on Remote Sensing, at St John's, Newfoundland.

Rossiter, J.R., and K.A. Gustajtis. 1978. Iceberg sounding of impulse radar. *Nature* 271:48–50.

Rossiter, J.R., and J.S. Holladay. 1994. Ice-thickness measurement. In *Remote Sensing of Sea Ice and Icebergs*, edited by S. Haykin, E.O. Lewis, R.K. Raney and J.R. Rossiter. New York: John Wiley & Sons.

Rossiter, J.R., and others. 1995. Remote sensing ice detection capabilities. Environmental Studies Research Funds.

Rothrock, D.A., Y. Yu, and G.A. Maykut. 1999. Thinning of Arctic sea ice cover. *Geophysical Research Letters* 26 (23):3469–3472.

Rott, H. 1984. The analysis of backscattering properties from SAR data of mountain regions. *IEEE Journal of Oceanic Engineering* 9 (5):347–353.

———. 1994. Thematic studies in Alpine areas by means of polarimetric SAR and optical imagery. *Advances in Space Research* 14 (3):217–226.

Rott, H., and R.E. Davies. 1993. Multifrequency and polarimetric SAR observations on alpine glaciers. *Annals of Glaciology* 17:98–104.

Rott, H., R.E. Davis, and J. Dozier. 1992. Polarimetric and multifrequency SAR signatures of wet snow. Paper read at IGARSS '92, at Houston, Texas.

Rott, H., G. Domik, C. Mätzler, H. Miller, and K.G. Lenhart. 1985. Study on use and characteristics of SAR for land snow and ice applications. Innsbruck: Institut für Meteorologie und Geophysik, Universität Innsbruck.

Rott, H., D-M. Floricioiu, and A. Siegel. 1997. Polarimetric and interferometric analysis of SIR-C/X-SAR data for glacier research. In *Proceedings of the EARSeL Workshop Remote Sensing of Land Ice and Snow*, edited by S. Wunderle. Saint-Étienne, France: European Association of Remote Sensing Laboratories.

Rott, H., and C. Mätzler. 1987. Possibilities and limitations of synthetic aperture radar for snow and glacier surveying. *Annals of Glaciology* 9:195–199.

Rott, H., C. Mätzler, and D. Strobl. 1986. The potential of SAR in a snow and glacier monitoring system. Paper read at SAR Applications Workshop, at Frascati, Italy.

Rott, H., and T. Nagler. 1992. Snow and glacier investigations by ERS-1 SAR — first results. Paper read at First ERS-1 Symposium: Space at the Service of Our Environment, at Cannes.

———. 1993. Capabilities of ERS-1 SAR for snow and glacier monitoring in Alpine areas. Paper read at 2nd ERS-1 Symposium, at Hamburg.

———. 1995. Monitoring temporal dynamics of snowmelt with ERS-1 SAR. Paper read at International Geoscience and Remote Sensing Symposium, at Firenze, Italy.

Rott, H., T. Nagler, and D-M. Floricioiu. 1995. Snow and glacier parameters derived from single-channel and multi-parameter SAR. Paper read at Extraction de parametres biogeophysiques a partir des donnees RSO pour les applications terrestres, at Toulouse.

Rott, H., W. Rack, P. Skvarca, and H. de Angelis. 2002. Northern Larsen Ice Shelf, Antarctica: further retreat after collapse. *Annals of Glaciology* 34:277–282.

Rott, H., and K. Sturm. 1991. Microwave signature measurements of Antarctic and Alpine snow. Paper read at 11th EARSeL symposium. Europe: From Sea Level to Alpine Peaks, from Iceland to the Urals, at Graz, Austria.

Rott, H., K. Sturm, and H. Miller. 1992. Signatures of Antarctic firn by means of ERS-1 AMI and by field measurements. Paper read at First ERS-1 Symposium: Space at the Service of Our Environment, at Cannes.

———. 1993. Active and passive microwave signatures of Antarctic firn by means of field measurements and satellite data. *Annals of Glaciology* 17:337–343.

Running, S.W., J.B. Way, K.C. McDonald, J.S. Kimball, S. Frolking, A.R. Keyser, and others. 1999. Radar remote sensing proposed for monitoring freeze–thaw transitions in boreal regions. *EOS Transactions* 80 (213):220–221.

Saich, P., W.G. Rees, and M. Borgeaud. 2001. Detecting pollution damage to forests in the Kola Peninsula using the ERS SAR. *Remote Sensing of Environment* 75:22–28.

Salisbury, J.W., D.M. D'Aria, and A. Wald. 1994. Measurements of thermal infrared spectral reflectance of frost, snow and ice. *Journal of Geophysical Research* 99 (B12):24234–24240.

Sandven, S., K. Kloster, and O.M. Johannessen. 1991. Remote sensing of icebergs in the Barents Sea during SIZEX 89. Paper read at First International Offshore and Polar Engineering Conference, at Edinburgh.

Sapiano, J., W. Harrison, and K. Echelmeyer. 1998. Elevation, volume and terminus changes of nine glaciers in North America. *Journal of Glaciology* 44 (146):119–135.

Saraf, A.K., J.L. Foster, P. Singh, and S. Tarafdar. 1999. Passive microwave data for snow-depth and snow-extent estimations in the Himalayan mountains. *International Journal of Remote Sensing* 20 (1):83–95.

Scambos, T.A., and R. Bindschadler. 1993. Complex ice stream flow revealed by sequential satellite imagery. *Annals of Glaciology* 17:177–182.

Scambos, T.A., M.J. Dutkiewicz, J.C. Wilson, and R.A. Bindschadler. 1992. Application of image cross-correlation to the measurement of glacier velocity using satellite image data. *Remote Sensing of Environment* 42:177–186.

Scambos, T.A., and M.A. Fahnestock. 1998. Improving digital elevation models over ice sheets using AVHRR-based photoclinometry. *Journal of Glaciology* 44:97–103.

Scambos, T.A., and T. Haran. 2002. An image-enhanced DEM of the Greenland ice sheet. *Annals of Glaciology* 34:291–298.

Scambos, T.A., C. Hulbe, M. Fahnestock, and J. Bohlander. 2000. The link between climate warming and break-up of ice shelves in the Antarctic Peninsula. *Journal of Glaciology* 46 (154):516–530.

Schaper, J., K. Seidel, and J. Martinec. 2000. Precision snow cover and glacier mapping for runoff modelling in a high alpine basin. Paper read at Remote Sensing and Hydrology, at Santa Fe.

Scherer, D., and M. Brun. 1997. Determination of the solar albedo of snow-covered regions in complex terrain. Paper read at EARSeL Workshop on Remote Sensing of Land Ice and Snow, at Freiburg.

Schmeltz, M., E. Rignot, and D. MacAyeal. 2002. Tidal flexure along ice-sheet margins: comparison of InSAR with an elastic-plate model. *Annals of Glaciology* 34:202–208.

Schmugge, T., T.T. Wilheit, P. Gloersen, et al. 1974. Microwave signatures of snow and freshwater ice. Paper read at Advanced Concepts and Techniques in the Study of Snow and Ice.

Schowengerdt, R.A. 1997. *Remote Sensing: Models and Methods for Image Processing*, 2nd ed. New York: Academic Press.

Seidel, K., C. Ehrler, J. Martinec, and O. Turpin. 1997. Derivation of statistical snow line from high resolution snow cover mapping. Paper read at EARSeL Workshop on Remote Sensing of Land Ice and Snow, at Freiburg.

Seidel, K., and J. Martinec. 1992. Operational snow cover mapping by satellites and real time runoff forecasting. Paper read at Snow and Glacier Hydrology, at Kathmandu.

Sellmann, P.V., J. Brown, R.I. Lewellen, H. McKim, and C. Merry. 1975. The classification and geomorphic implications of thaw lakes on the Arctic coastal plain, Alaska. Hanover, New Hampshire: U.S. Army Cold Regions Research and Engineering Laboratory.

Sellmann, P.V., W.F. Weeks, and W.J. Campbell. 1975. Use of side-looking airborne radar to determine lake depth on the Alaskan north slope. Hanover, New Hampshire: U.S. Army Cold Regions Research and Engineering Laboratory.

Semovski, S.V., N.Y. Mogilev, and P.P. Sherstyankin. 2000. Lake Baikal ice: analysis of AVHRR imagery and simulation of under-ice phytoplankton bloom. *Journal of Marine Systems* 27 (1–3):117–130.

Sephton, A.J., L.M.J. Brown, T.J. Macklin, K.C. Partington, N.J. Veck, and W.G. Rees. 1994. Segmentation of synthetic aperture radar imagery of sea ice. *International Journal of Remote Sensing* 15:803–825.

Serreze, M.C., J.E. Walsh, F.S. Chapin, T. Osterkamp, M. Dyurgerov, V. Romanovsky, W.C. Oechel, J. Morison, T. Zhang, and R.G. Barry. 2000. Observational evidence of recent change in the northern high-latitude environment. *Climate Change* 46:159–207.

Shashi Kumar, V., P.R. Paul, C.L.V. Ramana Rao, H. Haefner, and K. Seidel. 1992. Snowmelt runoff forecasting studies in Himalayan basins. Paper read at Snow and Glacier Hydrology. Proceedings published in 1993, at Kathmandu.

Sherjal, I., M. Fily, O. Grosjean, J. Lemorton, B. Lesaffre, Y. Page, and M. Gay. 1998. Microwave remote sensing of snow from a cable car at Chamonix in the French Alps. *IEEE Transactions on Geoscience and Remote Sensing* 36 (1):324–328.

Shi, J., and J. Dozier. 1993. Measurement of snow- and glacier-covered areas with single polarization SAR. *Annals of Glaciology* 17:72–76.

——. 1994. Estimating snow particle size using TM band 4. Paper read at International Geoscience and Remote Sensing Symposium. Surface and Atmospheric Remote Sensing: Technologies, Data Analysis and Interpretation, at Pasadena.

——. 1995. Inferring snow wetness using C-band data from SIR-C's polarimetric synthetic aperture radar. *IEEE Transactions on Geoscience and Remote Sensing* 33 (4):905–914.

——. 1997. Mapping seasonal snow with SIR-C/X-SAR in mountainous areas. *Remote Sensing of Environment* 59 (2):294–307.

Shi, J., J. Dozier, and R. Davis. 1990. Simulation of snow depth estimation from multi-frequency radar. Paper read at International Geoscience and Remote Sensing Symposium, at New York.

Shi, J., J. Dozier, and H. Rott. 1994. Active microwave measurements of snow cover: progress in polarimetric SAR. Paper read at International Geoscience and Remote Sensing Symposium. Surface and Atmospheric Remote Sensing: Technologies, Data Analysis and Interpretation, at Pasadena.

Shokr, M., B. Ramsay, and J.C. Falkingham. 1992. Preliminary evaluation of ERS-1 SAR data for operational use in the Canadian sea ice monitoring program. Paper read at First ERS-1 symposium: Space at the Service of Our Environment, at Cannes.

Shokr, M.E. 1991. Evaluation of second-order texture parameters for sea ice classification from radar images. *Journal of Geophysical Research* 96:10625–10640.

Shuman, C.A., R.B. Alley, and S. Anandakrishnan. 1993. Characterization of a hoar-development episode using SSM/I brightness temperatures in the vicinity of the GISP2 site, Greenland. *Annals of Glaciology* 17:183–188.

Shuman, C.A., R.B. Alley, S. Anandakrishnan, and C.R. Stearns. 1995a. An empirical technique for estimating near-surface air temperature trends in central Greenland from SSM/I brightness temperatures. *Remote Sensing of Environment* 51 (2):245–252.

Shuman, C.A., R.B. Alley, S. Anandakrishnan, J.W.C. White, P.M. Grootes, and C.R. Stearns. 1995b. Temperature and accumulation at the Greenland summit: Comparison of high-resolution isotope profiles and satellite passive microwave brightness temperature trends. *Journal of Geophysical Research* 100:9165–9177.

Shuman, C.A., and J.C. Comiso. 2002. In situ and satellite surface temperature records in Antarctica. *Annals of Glaciology* 34:113–120.

Sidjak, R.W., and R.D. Wheate. 1999. Glacier mapping of the Illecillewaet icefield, British Columbia, Canada, using Landsat TM and digital elevation data. *International Journal of Remote Sensing* 20 (2):273–284.

Singer, F.S., and R.W. Popham. 1963. Non-meteorological observations from satellites. *Astronautics and Aerospace Engineering* 1 (3):89–92.

Skriver, H. 1989. Extraction of sea ice parameters from synthetic aperture radar images. PhD, Electromagnetics Institute, Technical University of Denmark, Copenhagen.

Skvarca, P. 1994. Changes and surface features of the Larsen Ice Shelf, Antarctica, derived from Landsat and Kosmos mosaics. *Annals of Glaciology* 20:6–12.

Skvarca, P., H. Rott, and T. Nagler. 1995. Satellite imagery, a baseline for glacier variation study on James Ross Island, Antarctica. *Annals of Glaciology* 21:291–296.

Slater, M.T., D.R. Sloggett, W.G. Rees, and A. Steel. 1999. Potential operational multi-satellite sensor mapping of snow cover in maritime sub-polar regions. *International Journal of Remote Sensing* 20 (15):3019–3030.

Smith, B.E., N.F. Lord, and C.R. Bentley. 2002. Crevasse ages on the northern margin of Ice Stream C, West Antarctica. *Annals of Glaciology* 34:209–216.

Smith, D.M. 1998. Recent increase in the length of the melt season of perennial Arctic sea ice. *Geophysical Research Letters* 25:655–658.

Smith, F.M., C.F. Cooper, and E.G. Chapman. 1967. Measuring snow depths by aerial photogrammetry. Paper read at 35th Western Snow Conference, at Boise, Idaho.

Smith, L.C., R.R. Forster, B.L. Isacks, and D.K. Hall. 1997. Seasonal climatic forcings of alpine glaciers revealed with orbital synthetic aperture radar. *Journal of Glaciology* 43 (145):480–488.

Soh, L-K., and C. Tsatsoulis. 1999. Unsupervised segmentation of ERS and Radarsat sea ice images using multiresolution peak detection and aggregated population equalization. *International Journal of Remote Sensing* 20 (15):3087–3109.

Sohn, H-G., and K.C. Jezek. 1999. Mapping ice sheet margins from ERS-1 SAR and SPOT imagery. *International Journal of Remote Sensing* 20 (15):3201–3216.

Sohn, H-G., K.C. Jezek, and C.J. Van der Veen. 1998. Jakobshavn Glacier, West Greenland: Thirty years of spaceborne observations. *Geophysical Research Letters* 25 (14):2699–2702.

Solberg, R., and T. Andersen. 1994. An automatic system for operational snow-cover monitoring in the Norwegian mountain regions. Paper read at International Geoscience and Remote Sensing Symposium. Surface and Atmospheric Remote Sensing: Technologies, Data Analysis and Interpretation, at Pasadena.

Solberg, R., D. Hiltbrunner, J. Koskinen, T. Guneriussen, K. Rautiainen, and M. Hallikainen. 1997. Snow algorithms and products — review and recommendations for research and development. SNOWTOOLS WP410. Oslo: Norwegian Computing Centre.

Spring, W., T. Vinje, and H. Jensen. 1993. Iceberg and sea ice data obtained in the annual expeditions of the Barents Sea ice data acquisition program (IDAP). Paper read at POAC 93: Twelfth International Conference on Port and Ocean Engineering under Arctic Conditions, at Hamburg.

Stähli, M., J. Schaper, and A. Papritz. 2002. Towards a snow-depth distribution model in a heterogeneous subalpine forest using a Landsat TM image and an aerial photograph. *Annals of Glaciology* 34:65–70.

Stamnes, K., S-C. Tsay, W. Wiscombe, and K. Jayaweera. 1988. Numerically stable algorithm for discrete-ordinate-method radiative transfer in multiple scattering and emitting layered media. *Applied Optics* 27 (12):2502–2509.

——. 1995. Passive microwave studies of snow for the NOAA climate and global change program. Part 2. Bristol: Centre for Remote Sensing, University of Bristol.

Standley, A.P. 1997. The use of passive microwave and optical data in the SNOW-TOOLS project. Paper read at EARSeL Workshop on Remote Sensing of Land Ice and Snow, at Freiburg.

Standley, A.P., and E.C. Barrett. 1994. Passive microwave studies of snow for the NOAA climate and global change program. Part 1. Bristol: Centre for Remote Sensing, University of Bristol.

——. 1999. The use of coincident DMSP SSM/I and OLS satellite data to improve snow cover detection and discrimination. *International Journal of Remote Sensing* 20 (2):285–305.

Starosolszky, O., and I. Mayer. 1988. Characteristics of ice conditions based on aerial photography. Paper read at Ninth International Symposium on Ice, at Sapporo.

Steffen, K., R. Bindschadler, G. Casassa, J. Comiso, D. Eppler, F. Fetterer, J. Hawkins, J. Key, D. Rothrock, R. Thomas, R. Weaver, and R. Welch. 1993. Snow and ice applications of AVHRR in polar regions: report of a workshop held in Boulder, Colorado, 20 May 1992. *Annals of Glaciology* 17:1–16.

Steffen, K., J. Key, D.J. Cavalieri, J. Comiso, P. Gloersen, K. St Germain, and I. Rubinstein. 1992. The estimation of geophysical parameters using passive microwave algorithms. In *Microwave Remote Sensing of Sea Ice*, edited by F.D. Carsey. Washington DC: American Geophysical Union.

Steffen, K., and A. Schweiger. 1991. NASA Team Algorithm for sea ice concentration retrieval from Defense Meteorological Satellite Program Special Sensor Microwave Imager: Comparison with Landsat satellite imagery. *Journal of Geophysical Research* 96:21971–21987.

Stiles, W.H., and F.T. Ulaby. 1980a. The active and passive microwave response to snow parameters. 1. Wetness. *Journal of Geophysical Research* 85 (C2):1037–1044.

——. 1980b. Radar observations of snowpacks. *NASA Conference Publications* 2153:131–146.

Stouffer, R., S. Manabe, and K. Bryan. 1989. Interhemipsheric asymmetry in climate response to a gradual increase of atmospheric CO_2. *Nature* 342:660–682.

Stroeve, J., M. Haefliger, and K. Steffen. 1996. Surface temperature from ERS-1 ATSR infrared thermal satellite data in polar regions. *Journal of Applied Meteorology* 35 (8):1231–1239.

Strozzi, T., U. Wegmuller, and C. Mätzler. 1999. Mapping wet snowcovers with SAR interferometry. *International Journal of Remote Sensing* 20 (12):2395–2403.

Sturm, M., T.C. Grenfell, and D.K. Perovich. 1993. Passive microwave measurements of tundra and taiga snow covers in Alaska, USA. *Annals of Glaciology* 17:125–130.

Sugden, D.E., and B.S. John. 1976. *Glaciers and Landscape*. New York: John Wiley.

Sun, Y., A. Carlström, and J. Askne. 1992. SAR image classification of ice in the Gulf of Bothnia. *International Journal of Remote Sensing* 13:2489–2514.

Surdyk, S. 2002. Low microwave brightness temperatures in central Antarctica: observed features and implications. *Annals of Glaciology* 34:134–140.

Surdyk, S, and M. Fily. 1993. Comparison of microwave spectral signature of the Antarctic ice sheet with traverse ground data. *Annals of Glaciology* 17:337–343.

Svendsen, E., K. Kloster, B. Farrelly, O.M. Johannessen, J.A. Johannessen, W.J. Campbell, P. Gloersen, D. Cavalieri, and C. Matzler. 1983. Norwegian Remote-sensing experiment — evaluation of the Nimbus-7 scanning multi-channel microwave radiometer for sea ice research. *Journal of Geophysical Research — Oceans and Atmospheres* 88 (NC5):2781–2791.

Swamy, A.N., and P.A. Brivio. 1996. Hydrological modelling of snowmelt in the Italian Alps using visible and infrared remote sensing. *International Journal of Remote Sensing* 17 (16):3169–3188.

Swift, C., R.F. Harrington, and F. Thornton. 1980. Airborne microwave radiometer remote sensing of lake ice. Paper read at IEEE Electronics and Aerospace Convention.

Swithinbank, C. 1985. A distant look at the cryosphere. *Advances in Space Research* 5 (6):263–274.

———. 1988. *Satellite Image Atlas of Glaciers of the World: Antarctica*. Edited by R.S. Williams and J.G. Ferrigno. Vol. 1386-B, *U.S. Geological Survey Professional Papers*. Washington, DC: U.S. Government Printing Office.

Swithinbank, C.W.M., E.P. McClain, and P. Little. 1977. Drift tracks of Antarctic icebergs. *Polar Record* 18 (116):495–501.

Tait, A. 1998. Estimation of snow water equivalent using passive microwave radiation data. *Remote Sensing of Environment* 64 (3):286–291.

Tanikawa, T., T. Aoki, and F. Nishio. 2002. Remote sensing of snow grain-size and impurities from Airborne Multispectral Scanner data using a snow bidirectional reflectance distribution function model. *Annals of Glaciology* 34:74–80.

Tanré, D., C. Deroo, P. Duhaut, M. Herman, J.J. Morcrette, J. Perbos, and P.Y. Deschamps. 1990. Description of a computer code to simulate the satellite signal in the solar spectrum: the 5S code. *International Journal of Remote Sensing* 11 (4):659–668.

Tchernia, P. 1974. Étude de la dérive antarctique Est-Ouest au moyen d'icebergs suivis par la satellite Éole. *Comptes Rendus Hebdomadaires des Séances de l'Académie des Sciences* B278 (14):667–670.

Tchernia, P., and P.F. Jeanin. 1982. Some aspects of the Antarctic ocean circulation revealed by satellite tracking of icebergs. *Iceberg Research* 2:4–5.

Thomas, A., and D.G. Barber. 1998. On the use of multi-year ice ERS-1 σ^0 as a proxy indicator of melt period sea ice albedo. *International Journal of Remote Sensing* 19 (14):2807–2821.

Thomas, R., W. Krabill, E. Frederick, and K. Jezek. 1995. Thickening of Jakobshavns Isbrae, West Greenland, measured by airborne laser altimetry. *Annals of Glaciology* 21:259–262.

Thomas, R.H. 1993. Ice sheets. In *Atlas of Satellite Observations Related to Global Change*, edited by R.J. Gurney, J.L. Foster and C.L. Parkinson. Cambridge: Cambridge University Press.

Tinga, W.R., W.A.G. Voss, and D.F. Blossey. 1973. Generalized approach to multiphase dielectric mixture theory. *Journal of Applied Physics* 44:3897–3902.

Tsang, L., Z.X. Chen, S. Oh, R.J. Marks, and A.T.C. Chang. 1992. Inversion of snow parameters from passive microwave remote-sensing measurements by a neural network trained with a multiple-scattering model. *IEEE Transactions on Geoscience and Remote Sensing* 30 (5):1015–1024.

Tucker, W.B., D.K. Perovich, A.J. Gow, W.F. Weeks, and M.R. Drinkwater. 1992. Physical properties of sea ice relevant to remote sensing. In *Microwave Remote Sensing of Sea Ice*, edited by F.D. Carsey. Washington, DC: American Geophysical Union.

Ulaby, F.T., R.K. Moore, and A.K. Fung. 1982. *Microwave Remote Sensing — Active and Passive. Volume 2: Radar remote sensing and surface scattering and emission theory*. Reading, Massachusetts: Addison-Wesley.

Ulaby, F.T., R.K. Moore, and A.K. Fung. 1986. *Microwave Remote Sensing — Active and Passive. Volume 3: From theory to applications*. Reading, Massachusetts: Artech House.

Ulaby, F.T., W.H. Stiles, and M. Abdelrazik. 1984. Snowcover influence on backscattering from terrain. *IEEE Transactions on Geoscience and Remote Sensing* GE22:126–133.

Ulander, L.M.H. 1991. Radar remote sensing of sea ice: Measurements and theory. In *Technical report*. Göteborg: Chalmers University of Technology.

Unesco/IAHS/WMO. 1970. Seasonal snow cover. In *Technical Papers in Hydrology*. Paris: Unesco/IAHS/WMO.

Unwin, B., and D. Wingham. 1997. Topography and dynamics of Austfonna, Nordaustlandet, Svalbard, from SAR interferometry. *Annals of Glaciology* 24:402–408.

van de Wal, R.S.W., J. Oerlemans, and J. van der Hage. 1992. A study of ablation variations on the tongue of Hinteresiferner, Austrian Alps. *Journal of Glaciology* 38 (130):319–324.

van der Veen, C.J., and K.C. Jezek. 1993. Seasonal variations in brightness temperature for central Antarctica. *Annals of Glaciology* 17:300–306.

Vaughan, D.G., J.L. Bamber, M.B. Giovinetto, J. Russell, and A.P.R. Cooper. 1999. Reassessment of net surface mass balance in Antarctica. *Journal of Climate* 12 (4):933–946.

Vefsnmo, S., S.M. Løvas, S. Løset, and T. Næss. 1989. Identification and volume estimation of icebergs by remote sensing in the Barents Sea. Paper read at IGARSS 89: Twelfth Canadian Symposium on Remote Sensing. Quantitative Remote Sensing: an Economic Tool for the Nineties, at Vancouver.

Venkatesh, S., B. Sanderson, and M.S.S. El-Tahan. 1990. Optimum deployment of satellite-tracked drifters to support iceberg drift forecasting. *Cold Regions Science and Technology* 18 (2):117–131.

Vermote, E.F., and A. Vermeulen. 1999. *Atmospheric Correction Algorithm: Spectral Reflectances MOD09*. U.S. National Aeronautics and Space Administration.

(Cited 2003). Available from http://modis.gsfc.nasa.gov/MODIS/Data/ATBDs/atbd.mod08.pdf.

Vesecky, J.F., R. Samadani, M.P. Smith, J.M. Daida, and R.N. Bracewell. 1988. Observations of sea-ice dynamics using synthetic aperture radar images: automated analysis. *IEEE Transactions on Geoscience and Remote Sensing* GE26 (1):38–48.

Vesecky, J.F., M.P. Smith, and R. Samadani. 1990. Extraction of lead and ridge characteristics from SAR images of sea ice. *IEEE Transactions on Geoscience and Remote Sensing* GE28 (4):740–744.

Vikhamar, D., and R. Solberg. 2000. A method for snow-cover mapping in forest by optical remote sensing methods. Paper read at EARSeL Specialist Workshop on Remote Sensing of Land and Snow, Proceedings published in 2001, at Dresden.

Vinje, T. 1989. Icebergs in the Barents Sea. Paper read at Eighth International Conference on Offshore Mechanics and Arctic Engineering, at The Hague.

Vogel, S.W. 2002. Usage of high-resolution Landsat-7 band 8 for single-band snow-cover classification. *Annals of Glaciology* 34:53–57.

Vornberger, P.L., and R.A. Bindschadler. 1992. Multi-spectral analysis of ice sheets using coregistered SAR and TM imagery. *International Journal of Remote Sensing* 13:637–645.

Voss, S., G. Heygster, and R. Ezraty. 2003. Improving sea ice type discrimination by the simultaneous use of SSM/I and scatterometer data. *Polar Research* 22:35–42.

Wadhams, P., and J.C. Comiso. 1992. The ice thickness distribution inferred using remote sensing techniques. In *Microwave Remote Sensing of Sea Ice*, edited by F.D. Carsey. Washington DC: American Geophysical Union.

Wadhams, P., and N.R. Davis. 2000. Further evidence of sea ice thinning in the Arctic Ocean. *Geophysical Research Letters* 27 (24):3973–3976.

Wadhams, P., W.B. Tucker, W.B. Krabill, R.N. Swift, J.C. Comiso, and N.R. Davis. 1992. Relationship between sea ice freeboard and draft in the Arctic Basin, and implications for thickness monitoring. *Journal of Geophysical Research* 97 (C12):20325–20334.

Wakabayashi, H., M.O. Jeffries, and W.F. Weeks. 1992. C-band backscatter from ice on shallow tundra lakes: observations and modelling. Paper read at First ERS-1 Symposium: Space at the Service of Our Environment, at Cannes.

Walker, A.E., and B.E. Goodison. 1993. Discrimination of a wet snowcover using passive microwave satellite data. *Annals of Glaciology* 17:307–311.

Walker, A.E., and A. Silis. 2002. Snow-cover variations over the Mackenzie River basin, Canada, derived from SSM/I passive-microwave satellite data. *Annals of Glaciology* 34:8–14.

Wang, J., and W. Li. 2001. Establishing snowmelt runoff simulating model using remote sensing data and GIS in the west of China. *International Journal of Remote Sensing* 22 (17):3267–3274.

Wang, S.L., H.J. Jin, S. Li, and L. Zhao. 2000. Permafrost degradation on the Qinghai-Tibet Plateau and its environmental impacts. *Permafrost and Periglacial Processes* 11:43–53.

Warren, C.R., and D.E. Sugden. 1993. The Patagonian icefields: a glaciological review. *Arctic and Alpine Research* 25 (4):316–331.

Warren, S.G. 1982. Optical properties of snow. *Reviews of Geophysics and Space Physics* 20 (1):67–89.

———. 1984. Optical constants of ice from the ultraviolet to the microwave. *Applied Optics* 23:1026–1225.

Warren, S.G., and W.J. Wiscombe. 1980. A model for the spectral albedo of snow. II. Snow containing atmospheric aerosols. *Journal of Atmospheric Science* 37 (12):2734–2745.

Washburn, A.L. 1980. Permafrost features as evidence of climate change. *Earth-Sciences Review* 15:327–402.

Watanabe, O. 1978. Distribution of surface features of snow cover in Mizuho plateau. *Memoirs of the National Institute for Polar Research* Special issue 7:154–181.

Watkins, A.B., and I. Simmonds. 2000. Current trends in Antarctic sea ice: the 1990s impact on a short climatology. *Journal of Climate* 13 (24):4441–4451.

Weeks, W.F., and S.F. Ackley. 1986. The growth, structure and properties of sea ice. In *The Geophysics of Sea Ice*, edited by N. Untersteiner. New York: Plenum Press.

Weeks, W.F., A.G. Fountain, M.L. Bryan, and C. Elachi. 1978. Differences in radar return from ice-covered North Slope lakes. *Journal of Geophysical Research* 83:4069–4073.

Weeks, W.F., P.V. Sellmann, and W.J. Campbell. 1977. Interesting features of radar imagery of ice-covered North Slope lakes. *Journal of Glaciology* 18:129–136.

Weidick, A. 1995. *Satellite Image Atlas of Glaciers of the World: Greenland*. Edited by R.S. Williams and J.G. Ferrigno. Vol. 1386, *U.S. Geological Survey Professional Papers*. Washington DC: U.S. Government Printing Office.

Welch, H. 1991. Comparisons between lakes and seas during the Arctic winter. *Arctic and Alpine Research* 23 (1):11–23.

Welch, H.E., J.A. Legault, and M.A. Bergmann. 1987. Effects of snow and ice on the annual cycles of heat and light in Saqvaqjuac Lakes. *Canadian Journal of Fisheries and Aquatic Sciences* 44:1451–1461.

Wendler, G., K. Ahlnäs, and C.S. Lingle. 1996. On Mertz and Ninnis Glaciers, East Antarctica. *Journal of Glaciology* 42 (142):447–453.

Wessels, R.L., J.S. Kargel, and H.H. Kieffer. 2002. ASTER measurement of supraglacial lakes in the Mount Everest region of the Himalaya. *Annals of Glaciology* 34:399–408.

Whillans, I.M., and S.J. Johnsen. 1983. Longitudinal variations in glacial flow: theory and test using data from the Byrd Station strain network, Antarctica. *Journal of Glaciology* 29 (101):78–97.

Whillans, I.M., and Y. Tseng. 1995. Automatic tracking of crevasses on satellite images. *Cold Regions Science and Technology* 23:201–214.

Wiesnet, D.R. 1979. Satellite studies of fresh-water ice movement on Lake Erie. *Journal of Glaciology* 24:415–426.

Wildey, R.L. 1975. Generalized photoclimometry for Mariner 9. *Icarus* 25:613–626.

Williams, P.J., and M.W. Smith. 1989. *The Frozen Earth: Fundamentals of Geocryology*. Cambridge: Cambridge University Press.

Williams, R.N., W.G. Rees, and N.W. Young. 1999. A technique for the identification and analysis of icebergs in synthetic aperture radar images of Antarctica. *International Journal of Remote Sensing* 20 (15):3183–3199.

Williams, R.N., K.J. Michael, S. Pendlebury, and P. Crowther. 2002. An automated image analysis system for detecting sea-ice concentration and cloud cover from AVHRR images of the Antarctic. *International Journal of Remote Sensing* 23 (4):611–625.

Williams, R.S., J.G. Ferrigno, C. Swithinbank, B.K. Lucchitta, and B.A. Seekins. 1995. Coastal-change and glaciological maps of Antarctica. *Annals of Glaciology* 21:284–290.

Williams, R.S., and D.K. Hall. 1993. Glaciers. In *Atlas of Satellite Observations Related to Global Change*, edited by R.J. Gurney, J.L. Foster and C.L. Parkinson. Cambridge: Cambridge University Press.

Williams, R.S., D.K. Hall, and C.S. Benson. 1991. Analysis of glacier facies using satellite techniques. *Journal of Glaciology* 37 (125):120–128.

Willis, C.J., J.T. Macklin, K.C. Partington, K.C. Teleki, and W.G. Rees. 1996. Iceberg detection using ERS-1 synthetic aperture radar. *International Journal of Remote Sensing* 17 (9):1777–1795.

Wilson, L.L., L. Tsang, J.N. Hwang, and C.T. Chen. 1999. Mapping snow water equivalent by combining a spatially distributed snow hydrology model with passive microwave remote-sensing data. *IEEE Transactions on Geoscience and Remote Sensing* 37 (2):690–704.

Winebrenner, D.P., E.D. Nelson, R. Colony, and R.D. West. 1994. Observation of melt onset on multi-year Arctic sea ice using the ERS-1 synthetic aperture radar. *Journal of Geophysical Research* 99:22425–22441.

Wingham, D.J. 1995. The limiting resolution of ice-sheet elevations derived from pulse-limited satellite altimetry. *Journal of Glaciology* 41:413–422.

Wingham, D.J., A.L. Ridout, R. Scharroo, R.J. Arthern, and C.K. Shum. 1998. Antarctic elevation change 1992 to 1996. *Science* 282 (5388):456–458.

Winsor, P. 2001. Arctic sea ice thickness remained constant during the 1990s. *Geophysical Research Letters* 28 (6):1039–1041.

Winther, J-G. 1993. Landsat TM derived and in situ summer reflectance of glaciers in Svalbard. *Polar Research* 12 (1):37–55.

Winther, J-G., and D.K. Hall. 1999. Satellite-derived snow coverage related to hydropower production in Norway: present and future. *International Journal of Remote Sensing* 20 (15):2991–3008.

Wiscombe, W.J., and S.G. Warren. 1980. A model for the spectral albedo of snow. 1: pure snow. *Journal of Atmospheric Science* 37:2712–2733.

Wismann, V. 2000. Monitoring of seasonal snowmelt on Greenland with ERS scatterometer data. *IEEE Transactions on Geoscience and Remote Sensing* 38 (4):1821–1826.

Wunderle, S., and J. Schmidt. 1997. Comparison of interferograms using different DTM's — a case study of the Antarctic Peninsula. In *Proceedings of the EARSeL Workshop Remote Sensing of Land Ice and Snow*, edited by S. Wunderle. Saint-Étienne, France: European Association of Remote Sensing Laboratories.

Wynne, R.H., and T.M. Lillesand. 1993. Satellite observation of lake ice as a climate indicator — initial results from statewide monitoring in Wisconsin. *Photogrammetric Engineering and Remote Sensing* 59 (6):1023–1031.

Xiao, X., Z. Shen, and X. Qin. 2001. Assessing the potential of VEGETATION sensor data for mapping snow and ice cover: a normalized difference snow and ice index. *International Journal of Remote Sensing* 22 (13):2479–2487.

Xin, L., T. Koike, and C. Guodong. 2002. Retrieval of snow reflectance from Landsat data in rugged terrain. *Annals of Glaciology* 34:31–37.

Xu, H., J.O. Bailey, E.C. Barrett, and R.E.J. Kelly. 1993. Monitoring snow area and depth with integration of remote-sensing and GIS. *International Journal of Remote Sensing* 14 (17):3259–3268.

Yankielun, N.E., S.A. Arcone, and R.K. Crane. 1992. Thickness profiling of freshwater ice using millimeter-wave FM-CW radar. *IEEE Transactions on Geoscience and Remote Sensing* 30 (5):1094–1100.

Yankielun, N.E., M.G. Ferrick, and P.B. Weyrick. 1993. Development of an airborne millimeter-wave FM-CW radar for mapping river ice. *Canadian Journal of Civil Engineering* 20 (6):1057–1064.

Yi, D., and C.R. Bentley. 1994. Analysis of satellite radar altimeter return waveforms over the east Antarctic ice sheet. *Annals of Glaciology* 20:137–142.

Young, N.W., and G. Hyland. 2002. Velocity and strain rates derived from InSAR analysis of the Amery Ice Shelf, East Antarctica. *Annals of Glaciology* 34:228–234.

Zhang, T., R.G. Barry, K. Knowles, J.A. Heginbottom, and J. Brown. 1999. Statistics and characteristics of permafrost and ground ice distribution in the Northern Hemisphere. *Polar Geography* 23 (2):147–169.

Zhang, Y. 1999. *MODIS UCSB Emissivity Library*. Available from http://www.icess.-ucsb.edu/modis/EMIS/html/em.html.

Zibordi, G., and G.P. Meloni. 1991. Classification of Antarctic surfaces using AVHRR data — a multispectral approach. *Antarctic Science* 3 (3):333–338.

Zibordi, G., and M. van Woert. 1993. Antarctic sea ice mapping using the AVHRR. *Remote Sensing of Environment* 45:155–163.

Zwally, H.J. 1977. Microwave emissivity and accumulation rate of polar firn. *Journal of Glaciology* 18:195–215.

Zwally, H.J., M.A. Beckley, A.C. Brenner, and M.B. Giovinetto. 2002. Motion of ice-shelf fronts in Antarctica from slant-range analysis of radar altimeter data, 1978–98. *Annals of Glaciology* 34:255–262.

Zwally, H.J., and A.C. Brenner. 2001. The role of satellite radar altimetry in the study of ice sheet dynamics and mass balance. In *Satellite Altimetry and Earth Sciences*, edited by L.-L. Fu. New York: Academic Press.

Zwally, H.J., A.C. Brenner, J.A. Major, R.A. Bindschadler, and J.G. Marsh. 1989. Growth of Greenland Ice Sheet: Measurement. *Science* 246:1587–1589.

Zwally, H.J., and S. Fiegles. 1994. Extent and duration of Antarctic surface melt. *Journal of Glaciology* 40 (136):463–476.

Zwally, H.J., and M.B. Giovinetto. 1995. Accumulation in Antarctica and Greenland derived from passive-microwave data: A comparison with contoured complications. *Annals of Glaciology* 21:123–130.

Zwally, H.J., and P. Gloersen. 1977. Passive microwave images of the polar regions and research applications. *Polar Record* 18 (116):431–450.

Index

T - #0389 - 071024 - C16 - 234/156/14 - PB - 9780367392307 - Gloss Lamination